Elsdon Best, Department of Lands and Survey; New Zealand

Waikare-Moana, the Sea of the Rippling Waters

The Lake; the Land; the Legends. With a Tramp through Tuhoe land

.

Elsdon Best, Department of Lands and Survey; New Zealand

Waikare-Moana, the Sea of the Rippling Waters
The Lake; the Land; the Legends. With a Tramp through Tuhoe land

ISBN/EAN: 9783744753098

Printed in Europe, USA, Canada, Australia, Japan

Cover: Foto ©Andreas Hilbeck / pixelio.de

More available books at **www.hansebooks.com**

Department of Lands and Survey, New Zealand.

WAIKARE-MOANA,

THE SEA OF THE RIPPLING WATERS

THE LAKE; THE LAND; THE LEGENDS.

WITH

A TRAMP THROUGH TUHOE LAND.

BY

ELSDON BEST.

PUBLISHED BY DIRECTION OF
THE HON. JOHN McKENZIE, MINISTER OF LANDS.

WELLINGTON, NEW ZEALAND.
BY AUTHORITY: JOHN MACKAY, GOVERNMENT PRINTER.
1897.

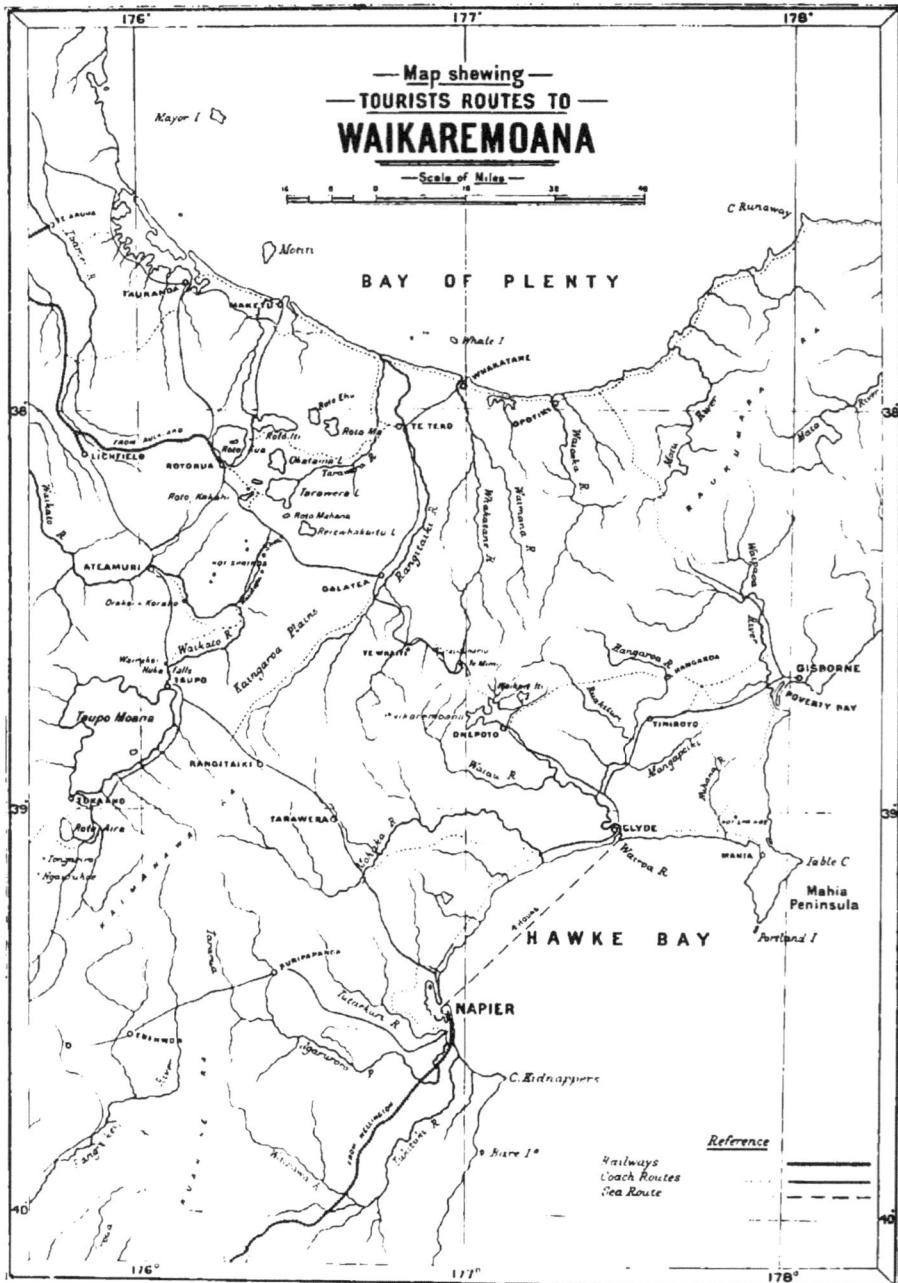

—Map shewing—
TOURISTS ROUTES TO
WAIKAREMOANA
—Scale of Miles—

BAY OF PLENTY

HAWKE BAY

Reference
Railways
Coach Routes
Sea Route

Department of Lands and Survey, New Zealand

WAIKARE-MOANA,

THE SEA OF THE RIPPLING WATERS:

THE LAKE; THE LAND; THE LEGENDS.

WITH

A TRAMP THROUGH TUHOE LAND.

BY

ELSDON BEST.

PUBLISHED BY DIRECTION OF
THE HON. JOHN McKENZIE, MINISTER OF LANDS.

WELLINGTON, NEW ZEALAND.
BY AUTHORITY: JOHN MACKAY, GOVERNMENT PRINTER.
1897.

GENERAL REFERENCE TO CONTENTS.

INDEX.

PREFACE.

LAKE Waikare-moana is situated not far from the east coast of the North Island of New Zealand, to the north-west of that great indentation named by Captain Cook Hawke Bay. The completion of the main road—in the near future—that leads from Rotorua to Gisborne will allow of a visit to this beautiful lake with something like ease. At the present time it must be approached from Napier as a starting-point. Small steamers cross Hawke Bay from Napier with tolerable regularity to the Wairoa River—a short trip of four hours—near the mouth of which is situated the pretty Town of Clyde. From Clyde a ride or drive of thirty-one miles along the road leading up the Wairoa and Waikare-taheke Rivers will bring the traveller to the outlet of the lake at Onepoto. From this point a road is in process of construction northward along the shores of the lake, to Aniwaniwa, the north-east extremity of the lake, and at which point a junction will be effected with the main road from Rotorua to Gisborne.

Of all the New Zealand lakes, Waikare-moana probably stands second for beauty, Mana-pouri taking the first place. It is often called the "Star Lake," from the number of arms which run far away into the hills, offering a series of most beautiful views of great variety. Everywhere the forest comes right down to the water's edge, whilst on the east side cliffs rise almost perpendicularly to close upon 2,000ft. above its surface. The lake offers delightful places for camping on the many beaches. Its height above sea-level is 2,050ft.

The following account has been printed by direction of the Hon. John McKenzie, Minister of Lands, with a view of furnishing information to tourists as to the various scenes of beauty on the lake : and at the same time an attempt has been made to invest the different places with a human interest by preserving the old Maori history relating thereto.

Young countries like New Zealand are often wanting in the historic interest associated with so many of the sights of Europe. This is not because New Zealand has no history, but because the guide-books fail to touch upon it. In the case of Waikare-moana, the isolation of the Maori inhabitants, until quite recently, has tended to preserve in the breasts of its people much more detail of the doings of the early occupiers than is usual, and this has been gathered together in the following pages. The whole of it is new matter, now for the first time collected from the Maori people themselves, and principally from the " Kaumatua," so frequently alluded to. This old man, whose name is Tu-taka-ngahau, is the hereditary chief of the Tama-kai-moana section of the Tuhoe Tribe, who by birth and education has the right to speak authoritatively on the history of his country. The Tuhoe tribes pride themselves on being the direct descendants of the aborigines whom the Maoris found here on the arrival of the fleet from Hawaiki in about the year 1350.

Thanks are due to T. Humphries, Esq., for most of the illustrations in this book.

S.P.S.

Waikare-Moana Road, near Te Whaiti.

WAIKARE-MOANA.

THE SEA OF THE RIPPLING WATERS.

WE were in camp at Te Whaiti-nui-a-Toi—the "Great Cañon of Toi "—in western Tuhoe land, when the word came to take the Rua-tahuna trail for Waikare-moana.

It was well timed, for the Kaumatua* was beginning to weary of the luxuries of life at the head camp, and to yearn for his beloved mountain solitudes, where the stern necessities of life suffice the hardy mountaineer, and the festive board is but indifferently well furnished ; while the Pakeha† looked forward with keen pleasure to viewing the mountainous region of the little-known parts of the famous Urewera country, the snow-wrapped peaks and mighty ranges, the vast forest and rushing torrents, the lone lakes and great gulches which form the leading features of Tuhoe land.   And, more-over, there comes to him, as there comes to all who truly love to view the face of mother Nature, the desire to look upon the un-wrought wilderness and note the war which has waged for untold centuries between it and primitive man—neolithic man, who has opened up the trails through the great forest he could not conquer— trails by which the incoming pioneers of the Age of Steel shall pass along, to leave behind them peace in place of war, thriving hamlets for stockaded pas,‡ fields of waving grain for jungle and for forest. And with this there also comes that strange sensation of vivid interest and pleasing anticipation which is felt by the ethnologist, botanist, and lover of primitive folk-lore when entering on a new field for research.   For the glamour of the wilderness is upon him, and the *kura huna*—the "concealed treasure" (of knowledge)— loometh large in the Land of Tuhoe.

A word here as to the Kaumatua, for methinks he is the leading character of this sketch.   An old man, probably sixty-five years of age, yet both strong and active, a leading chief of the Tuhoe Tribe, which has held these mountains for the last fifteen generations of

---

* Kaumatua, old man, a term of respect.   † Pakeha, a foreigner, a
white man.   ‡ Pa, a fortified village.

1

mixed blood, also a lineal descendant of the ancient race who dwelt in these lone lands long centuries before the present Maori people came across the great ocean from the isles of the sunlit sea. A warrior of the olden time, his face deeply scored by the chisel of the tattooer, possessing, moreover, a large-minded contempt for the habiliments of the white man. Here is a man who has faced death in many a fierce struggle, and led his hill-bred clan in many a gallant charge. And yet withal a quiet-mannered and courteous companion, ever ready to allay strife among his tribesmen, or to assist the stranger within his gates, be that stranger Pakeha or Maori. Such is the Kaumatau.

When the route came he said : " Friend! The word has come forth that you and I shall leave the parts trodden by the white man and go out into the lone places of the land, there to observe the homes of the old-time people — even unto the ' sea of the rippling waters,' which lies beyond the dark mountains of Huia-rau. It is well, O Son! I will be your guide through the great forest and across the snow-laden mountains, for I know well those rugged peaks and narrow passes, having trodden them many times. And my young men shall go with us to bear the heavy burdens, inasmuch as it is not wise to ascend among the snows of Huia-rau without good tents and much food. But do you keep in mind the ancient proverb. ' Ka haere te mata-tatahi, ka noho te mata-puputu.'* For truly am I waxing old, and the rough trails of my native land grow steeper year by year."

So that matter was settled, and, having sent forward by pack-horse the necessary supplies as far as Te Umu-roa, the Kaumatua and the Pakeha set forth by the new road now being formed from old Fort Galatea on the Rangi-taiki to Rua-tahuna, in the heart of Tuhoe land.

The scenery along this road is extremely picturesque and typical of the country. Leaving the Government camp at Wai-kotikoti, eighteen miles from Galatea and sixty-three from Rotorua, the road crosses the Whirinaki River at Rohutu, and winds up the hill to Niho-whati. For some distance here the road has cut through vast deposits of pumice, in which are seen the charred trunks of great trees destroyed by some great volcanic eruption in the long ago. Looking back from the hilltop we see the fine open valley of the Upper Whirinaki, bounded by great forest ranges. An historic district this, for here the " Multitude of the Marangaranga," the ancient people of the land, made a last futile stand against the conquering Maori, and the place teems with legends and quaint old stories. Upon the terrace below is Te Murumurunga, the village of the Ngati-Whare Tribe of aborigines, who are descended in part from the autochthones. On yon bluff above the surging river are seen the ancient walls of the Pa-o-Taketake, where that old warrior fought so well nine generations ago. To the right front

---

* This may be freely translated, " Fools rush in where angels dare not tread."

Ureweras, Chief Tutaka-Ngahau, his son Tukua-te-rangi, and his daughter-in-law.

is Te Harema, a palisaded fort on the hilltop, where a garrison of mixed tribes fought our Native contingent during the last war. The very spot we stand upon is an old battle-ground, where the descendants of Pukeko fell in the old pre-Pakeha days. Further up the valley are the ancient forts of the once powerful Ngati-Mahanga Tribe, who fell before the avenging spears of Tuhoe; while far away at the forest line across the upper valley is the Great Cañon of Toi, from which rugged chasm this district derives its name, and connected with which is many a strange legend of the days of yore.

We move on. The valley of the Okahu, a tributary of the Whiri-naki, is entered. Here the road is hewn out of the great rock bluffs which range steeply upwards, so that we look down through the tree-tops upon the rushing waters of Okahu far below. But a narrow gorge this Okahu, with high ranges on either side, covered with the far-reaching forest. Here is Ahi-manawa, so called from the fact that a chief named Tarewa-a-rua was here slain by his enemies, his heart torn out, cooked, and eaten by his delighted captors.

A warning cry comes from the "Children of Wharepakau," and we jump for the shelter of a protecting point. With a thunderous roar that makes the solid cliff tremble again a huge mass of rock leaps out from the bluff, and is hurled with a rattling crash upon the forest trees far below. The "children" have buried the war-axe, and taken to pick and bar and shovel. Oro-mai-take, another old Maori fort, is on the spur yonder, whence the warriors of Kihi fled from the men of Tawhaki after their futile attack on Te Hika Pa.

And so on, every hill and gulch and streamlet having its tale to tell, of war and battle and sudden death, in token of the "good old days." And now a changed and changing land is here, for behold! that fluttering fragment hard by beareth the fearsome legend "Old Judge," and the guileless sardine-tin lurketh by the wayside. An ominous sign, my masters!

The road winds up the range and through the dense bush until we look down upon the green forest lining the ravine of Manawa-o-hiwi. The peak of Tara-pounamu shows out, the summit of the great range which divides the watersheds of the Whirinaki and Whakatane Rivers. From this point a fine view of the surrounding country is obtained. Ranges, ranges, ranges! bush covered, lone, and silent, as far as the range of vision extends in every direction. Away before us looms giant Huia-rau, over which our way lies—a colossal range cutting the blue sky line, the pure snow glittering in the rays of a midday sun. Along the sierra on which we stand the peaks rise sharply up—Maro and Whakaipu and Te Pu-kiore, of which Te Arohana of old said, "*Kei pikitia a Te Pu-kiore.*"* But

---

* "Climb not the peak of Te Pu-kiore," but most probably meaning "desecrate not," &c.

it was, and by Hape of the Manawa Tribe, who fell at the taking of Oputara Pa, where the waters of Whirinaki rush forth from the gloomy gorge and ripple onwards through the plain of Kuha-waea.

But it is a far cry to Huia-rau, and so we leave the road of the white man and descend the steep range by the pack-trail, over which supplies are sent forward to the survey parties ahead of the road. A bush-track of the most primitive description winds down the rugged spurs and through the beautiful forest until we strike the Manga-pai Stream, a thousand feet below. Here is Te Wera-iti clearing, an old settlement of the Tuhoe, now long deserted. From here we travel down the channel of the stream through a wild forest gorge, with high ranges and rock bluffs on either side. After some miles of this mode of travelling, we leave the stream-bed and rise the terrace, emerging into the Native clearing at Te Umu-roa.

We are now in the heart of Tuhoe land, and within four miles of Mata-atua, the principal settlement of the Rua-tahuna district. Here reside the main body of the Tuhoe or Urewera Tribe, and here they have been for unknown centuries, for these are the lineal descendants of the ancient inhabitants of New Zealand. Century upon century have they held this mountain valley, ever keeping aloof from the tribes of the plain lands and of the coast, maintaining ever an aggressive attitude towards their neighbours, and holding in contempt those who could not trace their descent direct from Toi and Maru, and those deified ancestors who figure so largely in their ancient history. A strange people in a strange land, whose ancient system of *karakia* is most intricate and elaborate; who have preserved in these incantations hundreds of words from some archaic language of the shadowy past, and who are the remnant of a most ancient primitive race. And across the dark forest ranges which shoulder the rising sun, dwelling within the shadow of the sacred mountain Maunga-pohatu, are the remnant of Nga-Potiki—the "Children of the Mist," for are they not descended from Hine-pukohu-rangi, the Goddess, or Maid of the Mist? who lured to earth Te Maunga, or the Mountain, and whose issue was Potiki, whence comes the tribal name.

However, our route does not lie in that direction, and we camp at Te Umu-roa for the night. This is the furthest point to which a horse can be taken in the direction of Waikare-moana at present, and from here the route lies up the Rua-tahuna Stream for some miles until the Ngutu-wera Creek is crossed, whence the track winds up the range to the summit, at Te Whakairinga, near the Whakataka Peak.

So we are at Te Umu-roa, and preparing to camp for the night, when an offer is made by sundry young ladies of the Tuhoe clan to prepare for us our simple meal, or, as the sons of the Southern Cross put it, " to sling the billy." This offer we accept with cheer-

---

* *Karakia* —ritual, incantations, invocations, spells, charms, may all be included in this word as a general one, each division having its distinctive name.

Junction of Rua-tahuna and Whakatane Streams, Tuhoe Land.

ful alacrity, albeit we are well aware that these fair damsels have a keen eye to possible biscuits and cigarettes, two highly-prized luxuries in Tuhoe land. We then proceed to make ourselves comfortable in the *wharepuni*, or sleeping-house, where we are immediately surrounded by young and old, for the Pakeha is a new-comer in these parts, and is an object of curiosity to the primitive people of Rua-tahuna. The Kaumatua holds forth upon the subject of the outer world, and of the strange things he has seen in the camp of the white people at Te Whaiti-nui-a-Toi. And the Pakeha lights the pipe of peace and listens to the conversation going on around him, noting the different types to be seen among these people, and the singular nasal twang peculiar to the denizens of this district. Not that the latter sounds unpleasantly ; rather the reverse, the women speaking in a soft, low-toned drawl, which may be noted among the high-dwellers of Tennessee and other southern States. The subjects of conversation in these sleeping-houses appear trivial to a man of the outside world ; animated discussions are held anent the most minute details. This custom would seem to supply the place of written language to a primitive people, inasmuch as conversation supplies the place of literature.

And it was here in the *wharepuni* at Te Umu-roa, far away from those of his own race, and surrounded by the descendants of the unfortunate heroine, that the Pakeha first heard the sad story of Moetere and Houhiri, who died amid the snows of Huia-rau in the long ago. And as the tale applies to a certain place on our route we here relate it : —

How MOETERE AND HER LOVER PERISHED AMID THE SNOWS OF HUIA-RAU : A LEGEND OF THE GREAT SNOWY RANGE.

Manu-nui-taraki was a descendant of Tane-atua, who came to this land of Aotea-roa by the Mata-atua canoe (about the year 1350), and the generations from Manu-nui to the men of this time are eleven. Houhiri, son of Manu-nui, married Moetere, of the ancient tribe of Nga-Potiki, which tribe originally held all the lands from Maunga-pohatu to the waters of Tamahine-mataroa.* And these two dwelt at Tuku-roa, near unto Mata-atua, where they erected a house, and snared the birds of the forest, which were very numerous in the days of old. During one winter they resolved to ascend the Huia-rau Mountains in order to hunt the kiwi, for these great ranges are the home of the "hidden bird of Tane." They separated at Te Umu-roa, Moetere following the course of the Rua-tahuna Stream, which heads at Rua-tahuna Mountain, while Houhiri ascended the peak of Whakataka, where the flag of the Pakeha surveyor now flies. On arriving at the base of Huia-rau, Moetere followed the stream, which now bears her name, even to the summit of the great range. But while they were hunting the kiwi and kakapo in that rugged country a great snowstorm came

* The ancient name of the Whakatane River.

which lasted many days, and the snows piled high on Huia-rau. Then these forlorn people sought shelter from the fierce storm, and Houhiri crept into a rock shelter on far Whakataka, while Moetere, after vainly trying to make her way through the deep snow, gave up all hope of life, and laid down to die upon a rock which stands by the side of a small lakelet on the drear mountain. And so she died, and that rock has ever since been known as the Tapapatanga-o-Moetere. Died alone in that lone spot, while, on the gloom-hidden crest of Whakataka, her husband Houhiri, was chanting his death-song. And our love still goes forth to our ancestors who perished alone on the Great Snowy Range.

When Manu-nui heard that his children were lost to the world of light, he resolved to search for their remains, that he might take their bones to his home by the great ocean, that the sacred ceremonies pertaining to the dead might be performed over them, and that they might be laid away in the sacred place of his fathers.

So the old man journeyed to Rua-tahuna, and to Te Umu-roa, and to Te Mimi, where he entered the dark forest towards Whakataka. And as he went he murmured an ancient prayer of the Maori to enable him to find the bones of his son. Behold! by the power of that prayer did Manu-nui succeed in his quest, and the remains of his child were revealed to him by the gods of the ancient people. And the patriarch raised his voice in the wilderness and wept as he gathered the bones of his loved son, bleached by the snows of giant Huia-rau.

Then the heart of the old man went out to his daughter Moetere, and he traversed the rugged backbone of the *ika-whenua*[*] in search of her death-camp—that the bones of his children who loved each other so well might lie together, through the holy *pure*[†] in far away Whakatane. And as he went by low peaks and through the darkling woods he uttered the sacred *karakia*, which contracts or draws together the earth, for such were the powers and strange works of the men of old. Neither was it in vain, for it brought him to the dark pond where stands the lone rock, and on that rock lay the remains of Moetere and a few fragments of her clothing. Even so, O Pakeha! did Moetere and her lover perish on the great mountain, over which lies the trail to the Sea of Waikare, and before to-morrow's sun is lost behind the peak of Maro, you shall look upon the stream, which yet bears the name of Moetere, and camp amid the snows of Huia-rau.

And the childless old man went down through the silent forest to the low lands, bearing his sad burden to the shores of the Sea of Toi. When he came to Whakatane the sacred *pure* fire was kindled, and the cry of Manu-nui-taraki went forth: " O children! Here is food for the holy fire which gleams on Mou-tohora."

* Literally the " land-fish," the main backbone range of the country. The North Island is "the fish of Maui," hauled up by him from the ocean depths.
† *Pure*, purification.

The bones of our ancestors were then placed upon a stage, and a portion of the sacred food was given to the dead—that is, the *aria** of such food was absorbed by them, not the substance thereof.

Such is the story of Moetere, as related by her descendants in the *wharepuni* at Te Umu-roa. But it is now past midnight, and we must follow the example of the Kaumatua, and sleep that we may acquire strength.

CROSSING HUIA-RAU.

We were astir at daybreak on the following morning, and preparing for the day's march. The " children " of the Kaumatua are on hand, and soon reduce chaos to order in the way of making up the swags of tents, blankets, and rations. By this time a divine Hebe, in the person of Riri the *uru-kehu*,† appears with a huge " billy " of steaming tea, together with sundry and various viands of a non-luxurious nature. This trouble over, the carriers struggle into the swag-straps of their heavy burdens, and at a word from the Kaumatua go forward on their way. So, with the morning sun slanting down on Tahua-roa, and the voices of the Natives crying a farewell, we lift the Huia-rau trail for Waikare-moana.

The track for some distance led up a spur of the range which appears to head at Mount Rua-tahuna, from which peak the district derives its name, until we arrived at Te Mimi clearing, the site of a thriving settlement in former times, but which merely boasts of one lone inhabitant at the present time, a crippled old lady, who drags out some sort of an existence by the help of her descendants at Te Umu-roa. From here a fine view is obtained of the valley and district of Rua-tahuna. Far down the forest-shrouded valley, with Native clearings appearing at intervals, the range of vision is bounded by the hills closing in on the Whakatane River away below Mata-atua. Across the fern-ridge at Otekura is seen the roof of the great Council Hall of Tuhoe land, Te Whai-a-te-motu, which stands near unto the ruins of the more ancient one, Te Puhi-o-Mata-atua. To the right is Kiri-tahi, where, in the old fighting days, the Ngati-Porou contingent, under Major Ropata Wahawaha, built a pa and presented an aggressive shoulder to the wild and warlike bushmen of the Urewera. A fine view is here, looking down upon this primitive vale of Tuhoe land, untouched as yet by the Pakeha with his practical views of life.

The trail now descends into the Rua-tahuna Stream, and for miles we follow up its bed, cross and recross the rushing waters, and scramble along steep sidelings by the narrow track through brush- and forest- and fern-clothed gulches. And as we march, the Kaumatua discourses on places and incidents after the manner of his kind. For is not this the ancient war-trail of the Tuhoe Tribes? by

---

* *Aria,* essence, spirit, medium.
† *Uru-kehu,* light-haired : many of the Tuhoe people have light or reddish hair.

which they marched to attack the "Children of the Rising Sun," even from the days of Potiki and Ruapani of old; and by which, also, those same "children," with their perverted ideas of the rights of man, were wont to countermarch on the mountain hamlets of Tuhoe land in search of blood vengeance. How many a war-party has trodden this narrow track; fierce, tattooed warriors of the descendants of Awa, with their tribal priest skilled in the black arts, by which enemies are destroyed more surely than with club or war-axe!

And just here is a good example of the non-progressive barbarian, the conservatism of neolithic man. Here is an ancient highway between two districts—a path trodden by the Maori for full twenty-five generations—a path barely 6in. wide, and overhung with brush and ferns. Forsake it for a few months and the forest will obliterate it. It is like Mark Twain's house, inasmuch as it needs watching lest it be indistinguishable from the surrounding vegetation. Yet the Steel Age is here, and the stone *toki* (axe) is replaced by the products of Sheffield and Pittsburgh.

All these lands traversed by us from Te Mimi to Waikare-moana are now unoccupied of man, though the Kaumatau points out many places up the Rua-tahuna Stream, and on the western shores of Lake Waikare-moana, where the Urewera or Tuhoe people lived in bygone times. But the old-time *kaingas* (villages) are once again dense bush, and Te Whai-a-te-motu are limited to a few scattered hamlets at Maunga-pohatu and Wai-mana, and the vale of Whakatane.

We stop at one of these ancient settlements, known as Kapiti, to "sling the billy" for dinner, and the Kaumatua seats himself by the fire and relates the origin of this place name:—

"In olden times certain men of the Ngati-Ruapani, went to Lake Waikare-iti to snare the wily *parera* (ducks), and by the little isle of Te Kaha-a-tuwai they arranged a set of snares. The *kaha*, or line, was stretched across the water and fastened to a stake at each end. To this line the snares were attached in a long row, the loops being so arranged as to be suspended just above the surface of the water. Ere long a flock of ducks (*kawai parera*) passed through the channel and under the *kaha*, with the result that each snare—seventy in all—held a struggling duck. So stoutly did they struggle that their combined strength pulled up the stakes to which the *kaha* was secured, and the long string of birds, with snares and line and stakes, rose in flight, and in that manner flew as far as Kapiti, where the line became entangled in the branches of a huge kahika-tea-tree, and they were secured by the people of Tumata-whero. Hence this place became known as Kapiti, or Karapiti, which word signifies to be fastened in numbers side by side."

Here the Pakeha suggests that at least the stakes should be taken from the unhappy birds for their long flight over Huia-rau, but the Kaumatua holds stoutly on to those stakes. And what would you? for the kahika-tree still stands here, and Te Kaha-a-tuwai is yet known of man. But these few clearings, hewn out with stone axes and enlarged by means of fire, cannot hold the forest in check, and when abandoned are soon lost again in the surrounding bush. So much for the Stone Age.

We now cross the Ngutu-wera Stream, and stop a while at Pou-tutu, so named from the circumstance of a chief of that name belonging to the Ngati-Ruapani Tribe having been taken at this spot by the pursuing Tuhoe, of which more anon. On our left is a deep ravine wherein flows the Moetere Stream, of which we have seen the name origin, and up through the sombre tawai-trees comes the resounding roar of the falls. So far we have passed through a typical tawa bush, with rimu and tawai (*Fagus*) and ordinary undergrowth. We are now entering the higher regions, which are covered with a dense growth of tawai, tawari, and tawhero trees. The koareare shrub is here, the odorous leaves of which were woven into chaplets by the women of old, as also the tanguru-rake, which served a similar purpose. Further along, isolated on a peak of Huia-rau—to wit, Maunga-pohatu—are the kotara and pua-kaito, two rare and odoriferous shrubs, said by the Tuhoe people to be confined to that mountain. They were highly prized in former times, and were transplanted to the Native cultivations, though for some unknown reason it was considered an evil omen to transplant the kotara. For there were exquisites, mark you! in the days of yore among the warriors of Tuhoe land, and great pains were taken in the collection of sweet-scented leaves and herbs by the beaux of Rua-tahuna and Maunga-pohatu whereby to render themselves attractive to the fair-haired *uru-kehu* and the dark-browed daughters of Kuri. The oil of the titoki berries was scented with the gum of the tarata and the kopuru, a small plant found on rocks. In this oil was immersed the skin of a pukeko, or swamp-hen (*Porphyrio melanotus*), which was then formed into a ball and suspended from the neck, the skin resting on the wearer's breast. But when the missionaries of the Pakeha came they condemned this practice as savouring of the Evil One, and calculated to lead the Tuhoean soul to perdition.

On this great range are also many of the common shrubs and smaller trees, the kotukutuku, papauma, houhou, and raurekau, with the singular and beautiful neinei, and the toi or mountain palm. As for ferns, of a verity are we in the very heart of fern land among the gulches and cliffs of Huia-rau, and the heart of the Pakeha goes forth in love for these youngest and fairest children of Tane, the god of forests. For here are many acres by the trail-side covered with the beautiful punui (*Todea superba*), the reigning queen of ferns, and the graceful and feathery heruheru, the matata, and pipiko, and pakau-roharoha, and petipeti, and kawakawa, and many another, all interesting and all beautiful to those who will but look at them. Most common, however, here, as in many other districts, is the mauku, which, however, is not the less beautiful for being common. The young fronds, termed "*pikopiko*," formed an important article of food in the old pre-Pakeha days, and the matured fronds were rough-woven into coarse mats, used as clothing by the wild bushmen of these mountains of Tuhoe land. For in former times many of the interior *hapus* (or tribes) seldom saw the open country, but dwelt in the fastnesses of the rugged ranges. And, by the same token, the Tuhoe Tribes did not possess the better kinds of flax which make good clothing; they merely had the inferior kinds of a brash-fibre, such as grow on cliffs and hillsides. Hence this use of the mauku, and hence the old sayings, "*Rua-tahuna kakahu mauku,*" and "*Rua-tahuna paku kore.*"[*]

As we go onward many more varieties of shrubs are met with, the tapairu, and whinau-puka, and patu-tiketike, and tawheuwheu, and ngohungohu, and the kai-komako; which contains the sacred seed of fire according to ancient legend, for was it not to Hine-kai-komako that the primal fire fled in the days of Maui of the evil deeds, who deceived Mahuika, the goddess of fire? And to whom came Ira, with fair words and beguiling tongue, to whom was given the task of regaining fire for the sons of man. And the seeds of fire are still contained within the heart of Hine-kai-komako, and the generation thereof is well known by us. Here is the harsh tu-o-kura, from which the son of that famous warrior Te Kahu-o-te-rangi was named in the good old fighting days, and by yon stream is seen the hue-o-rau-kata-uri; on the cliff above is the trailing wae-kahu or *Lycopodium*. But the day weareth on apace, and he who lingers long by the wayside, of a verity shall he lay cold through the watches of the mountain night. After winding up the range above Ngutu-wera for some distance we arrive at Te Wharau-a-Te-Puia, which may be translated into the shelter of Te Puia. This gentleman was an ancester of the Ngati-Awa Tribe, who flourished some eight generations ago. When leading an expedition against the East Coast tribes he camped a night at this spot, and caused a shed (*wharau*) to be erected wherein he might pass the night—hence the name. Some distance further on we come to Te Wai-tuhi-a-Te-Ao-horomanga, a small place with an overgrown name, and where, doubtless, Te Ao, &c., tarried awhile to assuage a fine thirst induced by swagging his colossal cognomen over Huia-rau, inasmuch as *wai-tuhi* is a term applied to water which has collected in a hollow of a tree or log, and does not apply to water lying in pools on rocks or the ground. However, the tree containing this *wai-tuhi* has long since returned to mother Earth, and the next time Te Ao wends his weary way over Huia-rau it would be well were he to leave his name behind him, or bring a "billy" along.

At Te Pakura we strike the snow-line, and go in to camp for the night. The carriers cast down their heavy burdens with a sigh of relief, the tents are soon pitched, and the broad leaves of the toi palm collected for bedding. But the worst task is the kindling of a

---

[*] Rua-tahuna of the mauku garments. Rua-tahuna without property.

fire, no easy matter at this altitude in midwinter, for between rain
and mist and snow everything is wet and sodden to a painful
degree. It is fully an hour and a half before we can raise sufficient
fire to " sling the billy " on. It is accomplished, however, at last,
and the primitive " William " swings low to the unwilling flames
Supper over, we get under our blankets as quickly as possible, for
the cold is intense on the snow-line, the same being raw and damp
from the recent rains. The Pakeha regrets in mournful accents the
dry cold and huge fires of abertine and pitch-pine of olden camps, a
remembrance of the Rio Plumas and Sierra Nevada of the far north.
So we lay to rest on the rugged shoulder of rocky Huia-rau, being
literally *sub tegmine fagi*, and listen to the wail of Tawhiri-matea[*]
on the snowy peaks above.

There were no laggards next morning, for the cold roused us out
at daybreak, and we were soon mounting the steep ridge of Te
Pakura, which leads direct to the summit. Here we pass a fine
grove of the mountain palm (toi), with leaves 5in. wide ; and here
also is a small clearing, the first seen since leaving Te Mimi. Our
progress is by no means swift, for the swags are heavy and the Kaumatua must be considered, albeit the old man keeps sturdily though
slowly on the march. As we rise the summit we find the snow lies
deep, and has obliterated all signs of the trail. The summit is
covered with a dense growth of scrub, growing strong and close, a
typical chaparral, and no attempt has ever been made to cut a trail
through it, so that when a heavy fall of snow lies on the ground it
is somewhat difficult to keep the right track. The Kaumatua, however, never seems to be at fault, but trudges on barefooted, with a
serene indifference, through the ice-cold snow, dislodging heavy
masses of the same from the sturdy bushes as he pushes his way
through the thicket. We are fortunate, however, in having a fine
clear day to cross the summit, as would-be trans-Huia-rau travellers are often detained on account of heavy snow-storms at this
season, though the trip is a most enjoyable one in summer, when
one can dispense with a tent, or even blankets, for a few nights.
The brush being dense, we get no view of the lower country as we
traverse this lone and silent region until we suddenly break out on a
clear brow at Te Whaka-iringa-o-te-patu-a-Te-Uoro.

Here we rest awhile, for the name has tired us. Te Uoro was
a chieftain of Tuhoe, who flourished in the classic vales of Tuhoe land
some seven generations back, and as he was urging on his wild
career across Huia-rau one fair morn he encountered at this spot
one Te Amohanga, a lady of Ngati-Ruapani of that ilk, who dwelt by
the rippling waters of Waikare-moana. After some conversation on
the subject of tribal land rights, they decided that this place should
be the boundary-mark between the two tribes, on which Te Uoro
hung his weapon (*patu*) on a tree hard by as a sign of the compact.
Hence the above name, "The suspension of the weapon of Te Uoro."

* The God of winds.

This couple appear to have been well pleased with each other, for it is seen by a reference to tribal genealogies that the fair Amo was taken to wife by the warrior Te Uoro, though the weapon-suspending act would appear to have had but a transient effect, inasmuch as the aggressive sons of Tuhoe contracted a habit of shifting the weapon from tree to tree, and so the line crept further down Huia-rau year by year until the grim forts of Tuhoe rose one by one on the shores of the " Star Lake."

But it was worth the climb; for away below us lay the grand panorama of the lower country, the realm of the ancient Tauira, who held those lands by right from the god Maru of old, who held sway of Te Tini-o-Maru, far back in the very night of time. From the dark-blue waters of Waikare-moana, glistening in the trail of the sunlight, and gleaming between the wooded spurs 2,000ft. below : from the white cliffs of Pane-kiri to the massive range of Nga-moko, and far across the broken mass of ranges to the great bluffs which guard Kupe and Ngake, in the lone vale of Waiau, where nestle the lakelets of Te Putere, erstwhile the home of the banished " Children of Manawa," and sweeping northward across chaotic ridges, spurs, gulches, ranges, by the gloom-laden cañons of Hanga-roa and the silent caves of Tae : and past the sullen Reinga, with its old-time memories, where the roar of the great falls crashes through the darkling gulch as in the days of old when, five hundred years ago, the ill-fated and lovely Raka-hanga went down to death in that dark cañon : and drifting over the lonely lakes of Waihau, and the historic island fort which fell to the prowess of Tuhoe-potiki : to the bold peak of Whakapunake, the home of quaint legend : and Te Rau-o-piopio, where the mountain fortress of Rakiroa fell to the army of Ngapuhi, and Tuhoe and Tana-kakariki went down, and Te Ure-o-whata was abandoned of man, and Wai-reporepo was deleted from the roll of Kahungunu pas : then to the fair east-lands, and Te Whakaki-nui-a-Rua, with the great solemn ocean looking so near and yet so far away—the ocean of Kiwa of old—to far Te Mahia, where Waikawa breaks through the golden haze, seventy miles away. From white-browed Whakapunake to the dim Mataua-a-Maui, which looms afar off upon the horizon—the whole of this noble scene is spread out below us as we stand on the snow-wrapped crest of giant Huia-rau. Then the silver mist, lying low down upon the foot-hill, breaks, opens out, and drifts away up the dark gulches which have scarred the seared backbone of Tuhoe land from the days of Maui of old.

The bright inlets of the Star Lake run far up between the bush-clad ribs, which trend downwards to meet them from the mother range above. And over all the wondrous scene a great silence reigns ; the wind has died away, no sound comes from the voiceless forest, the rugged crags, the shimmering waters—silent, imposing, and grand - lies the untouched wilderness as upon the morning of the first day.

Even the talkative Native is silent, some looking upon the grand

scene for the first time, others scanning the lower lands to locate
some old-time camp, or the scene of some fierce struggle of the days
when they raised the war-axe against the invading white man.

And then the silence is broken by the mournful sound of a Native
tangi*. Standing alone upon the cliff brow, the Kaumatua rests
upon his staff and, looking down upon the well-remembered scene
below him, chants a long wailing lament for his old comrades who
have passed on to the Reinga; for his ancestors who dwelt and
fought here in the long ago; for the lands they paid for in blood,
and anguish, and much suffering. And listening to the lone old
warrior as he gives vent to his feelings in a strange, weird lament,
the Pakeha recognises the names of many an old-time hero of
Tuhoe land, of deeds long passed away, of fights fought long ago.

" Hail ! Ye lands of the rippling waters ; all hail, ye lands of our
ancestors of Tuhoe and Nga-Potiki. Hail to ye ! Children of the
mountain, whose bones lie beneath the dark waters, in the burial
caves of old, on many a hard-fought *parekura*.† It is you, O
ancient Hatiti, who fell at Te Maire there below, in lonely Whanganui. And you, O Toko! of the strong arm, who died as man should
die—in battle with upraised weapon. O helpless women and little
children ! whose bodies choked the Cave of Tikitiki—whose blood
reddened the waters of Wai-kotero—your bones have long since been
dust, but the hearts of Tuhoe still remember you. Rest you in
peace in your chamber of death, beneath the silent waters of Waikare, for the forest holds the crumbling walls of Nga-whakarara,
and from Te Ana-o-tawa, which darkens yon cliff at Ahi - titi,
methinks I yet hear across the waters the wail of Ruapani as we
drove them through the gates of death as *utu*‡ for your lives.
Greetings to you, O Children of the Mist ! for your *kaingas* (villages)
are silent and deserted and your lands trodden by a strange race.
No smoke rises around the silent sea, even from Te Mara-o-te-atua
to Te Korokoro-o-Tawhaki, and I alone of your generation am left—
I alone remain of the fighting men of old. Remain in peace, O
children ! for the strength I held to avenge you in days gone by
has now passed away, and the thought grows, that this is the last
time I shall climb this great *ika whenua*§ to greet you. *E noho
ra* ! "‖

As the old patriarch of the " People of the Mist " finished his
tangi for the dead of his tribe, he grasped his staff and strode
forward without a word. As silently the carriers take up their
burdens and move on after the Kaumatua.

Just below Te Whakairinga the trail descends a cliff and strikes
the head-waters of the Wai-horoi-hika Stream, misnamed by us as
Huia-rau. The snow upon the face of the cliff is frozen and
objectionably slippery, which renders our progress somewhat slow.
From this point the track simply follows the creek-bed for miles, at

* A lament for the dead.   † Battlefield.   ‡ Payment, revenge.
§ Backbone, main range.   ‖ Equivalent to farewell.

first over smooth rock, worn out by the erosive power of many waters until it resembles a huge trough. There are many falls in the course of the creek, some of which necessitate a detour by cliff or crevice. Lower down the bed of the stream is full of huge boulders, some of colossal size, which means much scrambling for the traveller. To those who have traversed this region it is somewhat amusing to be asked if horses can be taken to Waikare by the Huia-rau trail. The rocks here appear to be of three different kinds, *papa*,* shell-conglomerate, and sandstone. But little water flows in the stream-bed as we descend, though we note great drift-logs stranded on rocks 20ft. above its surface. This, together with the worn channel and bare banks, betoken what great floods must pour down this steep, rough gulch, when the rains are heavy and the snows are melting on the ranges above. The sight must indeed be a grand one at such a time.

As we fare on down the cañon the channel is enclosed by cliffs and steep ranges on either side. For some distance the hills are covered with a dense growth of scrub, the larger trees (tawai) having been killed by a bush fire years ago. Lower down we strike the bush again, and, looking down the rocky channel of Wai-horoi-hika, it presents a singular appearance, as the sombre beeches stand thickly on either side and their branches mingle overhead above the rushing stream. On the hill-sides are huge rocks and isolated masses, worn into strange forms by the weathering of many centuries, and high upon their soilless summits are gnarled and stunted tawai trees, which have sent long roots down the rugged rock-faces to seek nourishment far below. And so by cliff and fall and rugged ways we wend our way adown this mountain stream until within a mile of the lake, where the track leaves the creek and rises the spur, continuing down the top thereof. Here we pass through a fine open forest of tawai, tawhero, tawari, and toatoa or tanekaha, and also a good deal of neinei, which present a beautiful effect with their clean branches and tassel-like bunches of long narrow leaves. Very handsome sticks and canes are made of this shrub, for when the bark is removed the surface is found to be fluted in a singular manner ; also, as in the case of the toatoa, if the bark be removed by the agency of fire, the surface of the wood assumes a red colour, which same is highly esteemed by stick-collectors of Tuhoe land. On the brow of this spur, where the steep descent to the lake commences, is a little opening in the forest, from which we see the blue waters of the Whanganui Inlet lying beneath us—a charming scene as viewed through the trees, for the bush-covered hills trend abruptly down to the waters below. So we pass down the gully, where stand huge tawai of 6ft. and 7ft. diameter, and in a few minutes emerge into the world of light at Herehere-taua, the head of the Whanganui arm of the lake. Here we find the boat, lately placed upon the lake by the Government, with its Native crew

---

* A bluish-grey marl.

Near the Outlet, Waikaremoana.

waiting for us, and in a few minutes we are seated therein and pulling out across the placid waters of Waikare-whanaunga-kore[*] to One-poto, where the weary are at rest.

WAIKARE-MOANA.

But not for long. A bright and sunny morn finds us aboard our little craft, bent on the exploration of Waikare-moana. The tents and an ample supply of stores are stowed away, the Kaumatua takes his seat in the stern, as becomes the guide and philosopher of the party, the Native boatmen seize their oars, and we glide out of the little cove of One-poto and pass over the placid waters of the "Star Lake." Behind us rises the hill Rae-kahu, and between it and colossal Pane-kiri is a narrow pass or gulch, known as Te Upoko-o-te-ao, which faces the lake in the form of a steep cliff. This cliff was a famous *ahi-titi* in former times—that is, a place where the titi, or mutton-bird, was taken at night. This was done by means of a net, which was set up on the edge of the cliff, the net being braced or supported against poles, which were inserted in the ground and tied together at the top in the form of a triangle. The upper rope of the net was termed *tama-tane* (the son) and the lower one *tama-wahine* (the daughter) ; the X of the poles where lashed together above was called the *mata-tauira*. A fire was kindled on the extreme edge of the cliff in front of the net ; behind the fire, and immediately in front of the net, the titi-hunters seated themselves, each with a short stick in his hand for despatching the witless birds. Two men remained standing, their task being the killing of such birds as flew against the *mata-tauira*. Attracted by the fire, the titi flew against the net, where they were killed with a blow from the bird-hunter's stick. Should the first bird taken chance to fly against the *tama-tane* or *mata-tauira*, it was deemed an omen of ill-luck—the hunters would be unsuccessful *(puhore)*. Should the bird, however, strike the net at or near the *tama-wahine* —that is, near the ground—then the titi-hunters looked confidently for a good bag. A foggy night was selected for this important function, and great numbers of birds were thus taken in the old pre-Pakeha days, the same being highly prized as an article of food by the Maori. This industry is now, however, a thing of the past, for the European rat has driven the titi to the more remote and inaccessible parts. There were many such places around the lake, where this bird was formerly taken, another famous *ahi-titi* being the cliff near Te Wharawhara, and immediately above Te Ana-o-tawa. At Te Upoko-o-te-ao is seen the old redoubt where a detachment of the Armed Constabulary was stationed for many years. The crumbling walls of this relic of the war-times show plainly that the days of peace are here, and have " come to stay." Looking at this old defence, it is somewhat difficult to imagine what earthly or unearthly reason the builders thereof can have had to

* Waikare, relationless ; so called because its winds and waves are no respectors of persons.

build in such a position, for it is situated in a narrow saddle, with high hills on either side commanding it at short range. Fortunately, however, for the defenders, they were never attacked at this station.

We are now approaching the point known as Te Rahui, between which and Te Upoko-o-te-ao is Otau-rito. Te Rahui is a kind of meeting-place of the winds, and is much dreaded by Native canoemen when the lake is rough. The saying at such a time is " *Kia ata whakaputa i Te Rahui*"—that is, "Be careful in passing Te Rahui." If a canoe reaches Otau-rito safely when crossing in bad weather, the paddlers thereof consider that all danger is past. The *tohunga** of " Mata-atua," as the Native crew have named our craft, now commences his arduous task of initiating us into the ancient lore of Waikare-moana. Thus the Kaumatua: " The large, isolated rock you see at the point of Te Rahui is an ancient *whare pito tamariki*, or *takotoranga iho tamariki*, a spot where the *iho* (umbilical cord) of new-born children was placed as a *tohu whenua*. This custom, as it obtained in Tuhoe land, was to place the *iho* of children of succeeding generations at certain spots, in order to preserve the tribal influence over the lands adjacent. The *iho* was secured to a stone, and after the former decayed, the stone still maintained the name and power of that *iho*. This is an old custom, and I myself have seen it carried out. And across the lake, where you see the hill Ngaheni, at Opu-ruahine, there lies the *iho* of Hopa's brother, which preserves our *mana* over those lands. And it is from such dangerous places as Te Rahui that the lake derives its name of Waikare-whanaunga-kore. ' *Ka puta i Te Rahui, a ko te ao marama* '—' If you pass Te Rahui, you shall look upon the world of life.' "

We are now passing beneath the great Pane-kiri Bluff, which rises up 1,000ft. above us. This great cliff is one of the most imposing sights of this picturesque region, its white surface and bush-crowned summit being a striking landmark from many different points. The encroaching forest which meets the waters of the lake has assailed the bluff of Pane-kiri, and strips of hardy shrubs cling desperately to its rugged face, fill the narrow ravines and crevices, clamber along ledges, and finally, in several places, gain the towering crest far above. Frequently the softer strata of the perpendicular cliff have been weathered out, leaving a projecting ledge traceable for a long distance.

At Te Ara-whata is a steep ravine or cleft in the cliff face, where it is possible to ascend to the crest of Pane-kiri. This difficult ascent was often made in former times when the *kaingas* of Ngati-Ruapani were numerous on the lake-shores : hence the origin of the name.

Close to Ohiringi Bluff is a little cove, a good landing-place ; and here was situated one of the old Native settlements of years

---

* Wise man, expert, priest.

gone by. The old cultivations are grown up in scrub of many varieties; and at the base of Pane-kiri, which from this point trends off from the lake-shore, is the dark beech forest, mixed with rimu and miro. Looking out upon the lake from this point the scene before us is magnificent, for the waters of the lake, with a slight ripple thereon, are flashing in the rays of the morning sun; the green and beautiful forest sweeps up from the very water's edge to the peaks of the great ranges; the mass of Nga-moko stands boldly forth, while far away Manu-aha, snow-capped and rugged, looks clear and distinct across the lower ranges.

Past Te Papa-o-te-whakahu, a rock named after an ancestor of Ngati-Ruapani, who lived some ten generations back, we come to Tau-punipuni, where from the little inlet a noble view is obtained of the massive frontlet of Pane-kiri. The next little bay is Wai-tio, where a small stream runs into the lake, a stream famous for the number of pigeon-frequented trees which obtain near its source at the base of the cliff. Far above us we see the dark entrance to a cave, where doubtless the bones of many an old warrior lie, while far away westward across the Whare-ama Range there loom the great snowy mountains which stand above Waiau and Parahaki. Then on across the rippling waters to Wai-kopiro, another ancient settlement, with its wooded spurs and shrubs of many shades. At this place a small rivulet trickles down a rock-face into the lake, and these waters are said to possess some strange properties (*he wai kakara*, scented waters), for at certain seasons the little maehe fish come in myriads to drink these waters as they flow down the rock into the lake, at which times they are taken in great numbers by the Natives. This maehe, a small species of kokopu, is said to be the only fish in the lake, together with the koura, or fresh-water crayfish. Some Natives say that eels are also to be found, but that they have been introduced in late times from the Waikare-taheke River. Next comes Te Umu-titi, so named from the ovens (*umu*) used for cooking the mutton-bird (titi) which formerly abounded here. Then Paenga-rua Bay, a place noted for being windy; if the wind down this opening be strong no canoe can come out of Wairau-moana. The saying, "It is bad weather at Paenga-rua," is heard as far away as Rua-tahuna. Te Piripiri, is a famous spot among kaka (parrot) snarers, and Te Rawa, also a favourite resort of bird-catchers, the adjacent spurs of the Whare-ama Range being a famous *whenua pua*—that is to say, a land rich in the peculiar berries, and so forth, which the kaka, koko (or tui), and kereru (pigeon) feed upon. We are here informed by the Kaumatua that his tribe have a reserve at this place; doubtless a clear-headed people, these Tuhoe. At Te Rawa is a delightful little bay with a sandy beach, an ideal spot for the genus picnicker. Indeed, all around this inlet are many little coves and camping spots, the scenery being delightful; the bush slopes running back from the beach, and white cliffs visible at intervals through the dense forest growth.

2

We now head our craft round for the entrance to Wairau-moana, a long arm of the lake, which extends miles away to the west and south. As "Mata-atua" glides through the blue waters towards the narrow strait between Wairau and Waikare-moana, the glory of a gallant sun is upon the far forest ranges, and snowy peaks, the white cliffs to the far east are reflected in the clear waters of the lake, and far away across the silent waters are seen the blue cliffs which mark the approach to the land of "The Rainbow" (Te Aniwaniwa). At the point known as Te Horoinga the Kaumatua holds forth upon the local legend, which is to the effect that this point has the singular habit of changing its location, for it is said to recede before an approaching canoe, but remains stationary if the canoe stops—a habit doubtless that has been the primal cause of much aboriginal profanity. We note a reference to this belief in one of the local *waiata* or songs, "Ko te Horoinga e haere ana, e kore e tata mai" :—

> Tera te marama
> Tau whakawhiti rua mai,
> Kei runga ;
> Au ki raro nei noho noa ai ko au anake,
> Aroha ki te iwi ka nawaki ke atu ki tawhiti.
> Mokai ngakau,
> Ako noa au ki te mahi,
> Ka hua ai, a ko wai ?
> Ko te takakautanga i mua ra.
> Hua mai koutou e noho tikanga ianei,
> Tenei te tinana te whakapakia nei e te ngutu.
> Tu au ki runga ki nga haere a Te Riaki,
> Hei kawe i ahau,
> Arai kamaka ki Wha-koau.
> Au kia tu tonu he puna ngahuru,
> Nga kari noa,
> Koia ra nga tau i Te Horoinga.
> E haere ana e kore e tata mai.
> I te puke nui kei mate au.

> Behold the moon, there resting
> In its double path above,
> Whilst I alone am solitary below,
> Filled with love for the tribe so distant.
> In my despondent heart
> Vainly seek I some diversion.
> Methinks I am some other self.
> Had I but the freedom of yore !
> Thinkest thou that I am free from anguish
> Whilst this body is pierced by the lips' weapons?
> Would I could join with Te Riaki's company,
> And bear me far away,
> Beyond the screening rocks of Whakoau !
> But stand I like the springs in summer,
> Fruitlessly sinking, with vain striving,
> Like Te Horoinga of the song,
> Which passes onward, but is never reached.
> Let me not here die by the great hill's side.*

* All Maori poetry is acknowledged to be extremely difficult to translate ; indeed, to do so correctly requires the help of the composer. The above, and

Before entering the famous Strait of Manaia, we will take a look
back at Waikare-whanaunga-kore,* for we shall not see the main
lake again for some days.

This mountain-lake lies at an elevation of 2,050ft. above sea-
level, and the Huia-rau Range rises some 2,000ft. above the lake.
Waikare-moana is fed by many streams, the largest of which are
the Wai-horoi-hika, commonly called the Huia-rau Stream by Euro-
peans, the Opu-rua-hine, Mokau, Aniwaniwa, and Wai-o-paoa.
There is but one outlet, which is at Te Wha-ngaromanga, also known
as Te Wharawhara, close to Onepoto. The waters of this lake have
an eccentric habit of rising and falling as if endowed with tidal
power. This is due to heavy rains or melting snows, which cause
the lake to rise and overflow through the narrow rock-channel at
Te Wharawhara. Should the inflow from the many streams be
merely normal, the lake waters sink until the outlet-channel is dry,
and the only escape for the waters is by the subterranean passages
which are so numerous in the vicinity of the outlet. During our
visit the lake was at this low-water stage, and in traversing the
rough boulder-strewn beach from Te Kowhai Point to Te Ana-o-
tawa we could see in several places the waters rushing down between
the rocks, and hear the hoarse rumbling far below. At many places,
also, there are strong springs of water rising from below the bed of
the lake, and as we passed over them in the boat we could see the
rush of water issuing from the lake-bed and ascending with many
air bubbles to the surface. The outlet is a narrow passage some
12ft. in depth, cut by the waters through the solid rock, and is about
16ft. to 20ft. in width. When overflowing the lake waters rush
through this passage with great force, a tumbling mass of waters, in
which, as my informant tersely expressed it, " neither man, dog,
nor timber could live." At low water the underground outlets carry
the escaping water through the narrow rock-ridge to various points
some distance below the lake level, at which places it is seen
issuing from the hillside with tremendous force, and thence descends
the steep range in a series of cascades and foaming torrents to form
in the valley below the Waikare-taheke River.

On account of the broken nature of the country, Waikare-moana
is of somewhat singular form, there being so many inlets, bays, and
points. The Wairau branch, known as Wairau-moana, contains the
most beautiful scenery, for here are many little wooded islets, sandy
beaches, and small bays, with forest-covered points extending out
into the lake, the whole forming a most delightful and charming
scene. The surrounding forest contains many varieties of the most
beautiful ferns, and on the higher ranges are seen numerous rare

those to follow, are rough attempts to render into English something of the composer's meaning, but our language is wanting in many words to exactly express
those of the originals. The Maori is a poet by nature, and his poetry contains
many beautiful ideas when read in the original, which are universally marred in
the translation.—EDITOR.

* See ante, " Waikare, the relationless."

plants and shrubs. Within two hours' walk of the Whanganui-o-parua Inlet is the Waikare-iti Lake, a beautiful and little-known sheet of water, which lies some 500ft. higher than Waikare-moana.

It would be difficult to select a more delightful place in which to spend a holiday than the bays and inlets of the "Star Lake," as it is often termed on account of its shape, and the camper, artist, or geologist who would fail to enjoy such a holiday in Tuhoe land, let him camp by city streets, nor venture to lift the trail for Waikare-moana.

But we are now passing the narrow strait between Waikare and Wairau-moana, which is known as Te Kauanga-o-Manaia. This Manaia is said to have been a chief of the ancient tribe Te Tini-o-Tauira, and, having swam across this passage in those bygone times when his people held sway here, the strait has ever since been known by the above name. On our left is Nga-whatu-a-Tama, a small mound on a point of land jutting out into the lake, and connected with the mainland by a low, narrow neck. This mound was one of the ancient pas of the Ngati-Ruapani Tribe, by which they held this district. Hither the refugees from Whakaari fled when defeated by the sons of Tuhoe. Like all the old forts around the lake, it is now covered with a dense forest growth. It is said to have been named after Tama or Rongo-tama, another chieftain of the ancient Tauira Tribe. An historic spot this, as it guarded the entrance to Wairau-moana in the old fighting days, when the shores of the now lonely sea of Waikare, were covered with many cultivations, and men worked with weapons in their belts, and the many fighting pas were thronged with the children of Ruapani and Hine-kura, of Te Uira-i-waho and Parua-aute. And well might Tama of old watch the Pass of Manaia, for were not the ancient Nga-Potiki, the "Children of the Mist." who dwelt among the snows and cliffs of Maunga-pohatu, ever watching and waiting for an opportunity to attack the "People of the Rising Sun," who slew Hatiti, born of the "Mountain Maid"?

As we round the protruding "Eyes of Tama" the beautiful Inlet of Te Puna opens up to the west. The morning mist is rising from the glassy waters, the sun glitters and dances along the smooth surface and lights up the green forest, which meets the gleaming waters; the song of many birds comes from the hillsides and beautiful islets across the placid waters, the great ranges in the far distance bound the line of vision.

It is Ohine-kura, the place of many baylets and miniature isles, named from Hine-kura, an ancestress of Ruapani, slain by Tuhoe some ten generations ago. Here we are hailed with an old-time greeting by a son of the soil, Hurae Puketapu, of Ngati-Ruapani, the only human being encountered by us in our trip round the lake, and who is hunting the wild hog and shooting pigeons on the lands of his ancestors, occupying the intervals in hewing out a canoe which we opine will be ready to launch some time before the dawn of the twentieth century. And Hurae is evidently a hospitable fellow, for he invites us to land and partake of his forest fare, and then, recognising a Pakeha, he bids him welcome to Wairau and the fatness thereof, "for we are one people now." So we exchange greetings from the shining waters below and rocky cliff above, while the crew of "Mata-atua" fill the cheering pipe and watch the koura, or crayfish, on the sandy bottom 30ft. beneath her keel. So we fare on by point and bay and wooded isle to Korotipa, remarkable for the number of pretty little coves in its vicinity, and from which place the view, looking ahead up Wairau, is a sight for the gods, for the great encircling ranges in the background seem to give the lake a double beauty. Then the baby islet of Nga-whakarara, another old stronghold of the Ruapani people, and where they were defeated by Tuhoe and hunted far away towards the coast. And where—but the Kaumatua here goes out on strike, and says that the story of that fight is too long to relate now, but we will have it round the camp-fire at night, merely stopping to point out the spot where Tipihau, of Tuhoe, slew Pare-tawai during that sanguinary struggle.

Thence we come to Nga Makawe-o-Maahu. We are drifting back into the remote past now, and the *ao marama* (or world of light and being) is far behind us, inasmuch as the renowned Maahu had his being in the dim dawn of time when gods deigned to dwell on earth. For was it not he who engaged Haere, the rainbow god, in combat, what time the Tini-o-kauae-taheke people descended the sacred pohutukawa * before the divine sons of Houmea? And Hau-mapuhia, son of Maahu; who has not heard of his great feat in forming the Waikare-moana Lake in the misty days of yore? Maahu, of the mystic land, a name to conjure with on the classic shores of Wairau-moana! And here is Nga Makawe-o-Maahu, the hairs of his sacred head, represented by those plants of harakeke (native flax) growing on the cliff yonder. They are very sacred hairs as befits so great a man, and if they are touched or interfered with in any way, woe betide the luckless wight who so offends, for if the gods do not kill him they will cause him to remain the balance of his days in Wairau-moana, and be the waters never so calm, and paddle he never so bravely, yet shall it be in vain, and he who insults the hairs of Maahu shall never pass through Te Kauanga-o-Manaia, but spend his weary days in paddling ever towards Nga-Whatu-a-Tama, which he shall never reach.

Those singular round boulders on yon point are also named in honour of this famous ancestor. They are Nga Whanau-a-Maahu, the "Children of Maahu," who are probably awaiting the return of their erring parent from the great ocean of Kiwa. However, those "children" are by no means sacred, and you may go and look closely at them if you wish, or even at the sacred things of this land, for these laws do not possess *mana* (power, influence) over the Pakeha.

---

* The spirits of the dead descend over the cliff at the North Cape to Te Reinga, or Hades, by means of the roots of pohutukawa trees.

22

Past Te Ana-a-kakapu is the beautiful bay of Wha-kenepuru, a lovely spot, with a short sandy reach of shore-line, and the picturesque wooded isle of Te Ure-o-patae in the foreground. Across the calm waters of the bay a black swan* glides in a stately manner, followed by her young, wondering, no doubt, at this invasion of her lone domain.

One-tapu—the sacred strand of Maahu—where the Kaumatua tells us how the rebel leader Te Kooti, when retreating from Mohaka, brought a mob of horses through the back country to Te Wai-o-paoa, at the extreme south-west point of Wairau-moana, thence by the rugged shore to the sandy beach of One-tapu, where he and his band camped for some days amusing themselves by holding horse-races on the beach. From here Te Kooti took the horses as far as Nga Whatu-a-Tama, where he swam them over Te Kauanga-o-Manaia to the opposite shore, and then, ascending the rugged spurs of Huia-rau, managed to get some of his stud of stolen horses across that fearful country to Rua-tahuna, though many were killed during the journey. As we glide past Motu-ngarara, a bush-covered island on which yet another ancient Maori pa stands, we see a large flock of ducks paddling along the shore-front, and regret the absence of our gatlings.

At the promontory of Te Kaha, almost surrounded by water, we land and lunch, and, while the boatmen are elevating the sober "William" that cheers without inebriating, we will take a look back on Wairau-moana. For it is truly magnificent, with the little isles looking as groves of trees upon the face of the shining lake, and the sun flashing in the waters of many inlets; with the noble forest of Tane† sweeping back by ridge and range to colossal Huia-rau, with its covering of glittering snow, and Manu-aha, which pierces the distant sky-line. So the Kaumatua and the Pakeha look upon this most picturesque of mountain-lakes, and discourse anent the ancient history thereof and the wondrous tales of old—of wood-elves in the sombre forests, and fierce *taniwha* (demons, dragons, &c.)—in deep pools, of strange creatures among the great mountains, and goblins by cliff and cave—until the call to a frugal meal comes from the "children," and is promptly obeyed.

It is well that we have dined, for we are now approaching three most sacred places where it would be the blackest sacrilege to convey cooked food. These places are Te Pa-o-Maahu, where that *tupuna* (ancestor) was wont to reside; Te Wai-kotikoti-o-Maahu—the sacred spring of Maahu; and Te Puna-a-taupara, whence the Maahu household derived their water-supply for domestic purposes, and in which the ill-fated Hau-mapuhia came to an untimely end, and thereby acquired god-like powers. Te Pa-o-Maahu is a most picturesque little wooded knoll standing on a small flat at the head of the bay. Another relic of Maahu is

* Black swans were introduced from Australia many years ago.   † Tane, the god of forests, and birds.

Wairau-moana, looking N.E. from Te Ure-o-patae Island.

his sacred dog, an animal possessed of strange powers, and which lives beneath the waters of Te Roto-nui-a-ha, a small lake at Te Tapere, where are also two other lakes, known as Roto-ngaio and Roto-roa. The aforesaid dog has the faculty of *matakite* or prophecy, and is heard to bark beneath the waters of the lake whenever the death of a chief is near. At Te Putere also the remnants of Ngati-Manawa found a refuge when they fled from Te Waiwai and Tarawera, where they had retreated after the fall of Okarea Pa, on the Wai-a-tiu, a tributary of the Whirinaki, near unto Te Whaiti-nui-a-Toi. And though Tuhoe had assisted Ngati-Pukeko at the siege of Okarea in order to avenge the killing of Matua, Tai-mimiti, and Tuara-whati, yet they took the refugees of Manawa from Te Putere to Rua-tahuna, where they appear to have treated them well, with the exception of having put them in old *kumara* pits in lieu of houses, and to this day it is not meet to mention those pits to Ngati-Manawa. And, again, at Te Putere is a waterfall which flows over a cliff on which are two projecting rocks, called respectively Kupe and Ngake, though how the names of those most ancient heroes and navigators came to be located here is indeed difficult to say. However, they serve a useful turn, as when, in chasing eels down stream, the Natives drive them over the fall, they are dashed by the waters down on to the back of Kupe, whence they rebound on to Ngake, who indignantly casts them far out upon the bank below, where the wily Maori secures them.

But we must return to Te Wai-kotikoti-o-Maahu, which is the name of a spring of water, and a sacred place (*tuahu**) of Maahu of old, where the most sacred operation of hair-cutting was performed on his thrice sacred head. It was also a *wai-whakaika* of that ancient warrior, where, after the hair-cutting ceremony, he went through the rites of the *wai taua*, of which there are several, all attended with many sacred *karakia* (incantations) and due solemnity. The *tira* was one of these, a rite by which the sins and evil thoughts of the members of a war party were wiped out, and they went forth on the war trail with a clean sheet, prepared to serve the god of war, Tu, with faithful devotion. In this *tira* ceremony the *tohunga*, or priest, took off all his clothing and donned the *maro-huka*, the sacred girdle. In this scant attire he went to the *wai-whakaika*, where he formed two small mounds of earth, in each of which he placed a twig of the karamu tree, called a *tira*, or wand. One of these is the *tira ora*, or wand of life, and the mound of earth it rests in is the *tuahu-o-te-rangi* the (altar of heaven). The other is the *tira mate*, the wand of death, the mound being *puke-nui-o-papa* (the great hill of earth). By means of his potent *karakia* the priest causes the *tira mate* to absorb all the sins and evils of the members of the *taua* (war party) – that is, it is the *aria* (or medium) of those evils. The priest then dons his *tu-maro* (war-girdle) and proceeds to weaken

---

* Tuahu, a place where incantations were offered up and other rites performed : an altar, in fact, though unlike one in shape.

the tribal enemies by means of *makutu* (or witchcraft), which comprehends a vast series of prayers, incantations, and ceremonies, the final *karakia* being those named *maro* and *wetewete*.

Also at this holy spring was cut the hair of the *tauira* or students of the *wharekura*, a building where the ancient lore, genealogies, and history of the tribe were taught. At the completion of the lesson in *wharekura*—that is, at dawn of day—the priest led the scholars to the spring, where he cut the hair of each one with a flake of *mata* (obsidian), which rite was termed *wai kotikoti*. After this came the *wai-whakaika* and *wai-tana*, as described above.

But we must leave the sacred spring of Maahu and urge on, for the sun is hanging low on the ranges and we must camp betimes. We are now approaching the end of Wairau-moana, and the opposite shore trends in towards us as we advance. A lone rock with a single stunted tawai-tree growing thereon, the smallest of islets, lies 100 yards from the shore : it is Te Whata-kai-o-Maahu, where that old warrior was wont to store his food.

So "Mata-atua" is turned to the beach, and we land at Wai-o-paoa and pitch our tents on a little grassy flat, having hauled our good craft ashore. And while the "children" are fixing the camp and gathering fuel we will ascend the fern-ridge between the two streams, for a most beautiful view of the Wairau branch is obtained from that point. The lake lies far beneath, broken into innumerable inlets, with bushy islets and points ; the ranges shelve steeply down to the lake-shore ; the range of Whare-ama cuts off the view of the main lake, though great Pane-kiri is still in evidence. A great silence broods over the shining waters of Wairau ; the forest, the waters, the hills of this ancient abode of man are silent with the desolation of a passing race. The fighting pas of old lie numerous before us ; the lake-shores are covered with the sites of former cultivations, each hill and point, bay and isle carries its legend of the long ago, when the children of the soil were numerous in the land of the ancient people. No smoke arises in all this great expanse, no human beings but ourselves lay down to rest this night on the shores of Wairan-moana. *Kati !* Let us hurry back to camp that we may learn of Maahu, and Rua, and Maru of old before it is too late. For the lands of Waikare are in a transition stage—the Maori has gone, though the Pakeha has not yet arrived ; yet a little while and it will be too late.

Night settles down upon the silent lake, the cheerful camp-fire gleams brightly across the placid waters and lights up the white tents, a myriad brilliant stars are seen in the clear bosom of Wairau-moana, reflected from the clear sky above. The rime of white frost sparkles on sedge and rock, but the fire, built by cunning hands, is bright and warm, and the joy of the Bohemian mind is with us. Anon the white mist creeps down the sombre gulches and spreads out across the silver lake, obscuring isle, and mount, and rocky cliff.

The blankets are spread before the tent and facing the cheery

Wairau Arm, Waikaremoana, looking N.E. from Wai-o-paoa.

(From a sketch by S Percy Smith.)

log fire, and, with the beloved pipe, which softeneth the heart of man, we take in the beauty of the glorious scene before us, while the Kaumatua recites the tales of yore, the deeds of the god-like men of old, strange doings of monsters and semi-human creatures which lived in these weird places of the earth, before the Maori came across the dark ocean. The boat has gone away in care of the "children"; gone to One-poto, the parts trodden by the white man, and the Kaumatua and the Pakeha are left alone in the realm of Maahu, the lonest spot in lone Wairau. And then, with the *kaingas* (dwelling-places) of the ancient people around us, the scenes of the exploits of the ancestors of Tuhoe and Ruapani, the forts of the old-time tribes still vivid in the mind, alone in the great, silent expanse of Wairau-moana, the time has surely come to learn what is known of those who lived and fought and died in these mountain solitudes, long centuries before the white man dared adventure the great ocean of Kiwa.

The Kaumatua draws his blanket around him, his deeply-tattooed visage lighting up with interest, he extends his bare arm towards the lake, and the "Oracle of the Rocky Mountain"* speaks:—

"*E pai ana, E hoa!*† Now that Hine-pukohu-rangi is descending from her ancient love, our ancestor Te Maunga, whom she lured to earth in the days of long ago, and here among the silent homes of the ancient people, it is well that I should tell you the legends of the 'Sea of the Rippling Waters,' for that is why I followed you through the dark forests and across the snowy mountains which lie far away, where the sky hangs down. And it is not an idle journey, but one in which there is much to be learned and much to be seen. But do you not be alarmed at the monsters which inhabit this 'Sea of Waikare,' for I am an *ariki taniwha*,‡ I am descended from Rua-mano, and Nga-rangi-hangu, and Te Tahi-o-te-rangi, who were *taniwha* ancestors of mine, though some descended from the trees of the forest—that is, from the children of Tane-mahuta, such as the Te Marangaranga Tribe—therefore it is well that I should be with you, for no *taniwha* will molest me; and do you be strenuous in retaining what I impart, for I know that you have not eaten of the sacred herb which binds knowledge acquired. Remember the 'Ahi-o-pawhera' and the fate which overtook that ancestor of Tuhoe land. Friend, it is well that we are alone, for my children who go with us have little love for the gallant stories of old, and I will tell them to you and to one other and no more, that you may preserve these traditions of my people and record their ancient customs, that they may be retained in the world of light. And do you write them plainly in your *paipera*,§ that all who love such things may understand, for I would even hope that my children may yet return to the *kura*‖ of Tuhoe and of Potiki and be proud of the achievements of their ancestors. *Tena!*"

---

* Rocky Mountain, Maunga-pohatu, the Kaumatuas' ancestral home.   † It is meet, O Friend!   ‡ Lord of dragons.   § Bible: any large book used for recording is so called by the Maoris.   ‖ Knowledge, valuable possession.

THE ANCIENT PEOPLE OF WAIKARE-MOANA.

In the days of old, long before the time of Mura-kareke and Tuhoe-potiki—who lived eighteen generations ago—an ancient tribe, known as Te Tauira, held all these lands of Waikare and far away to Waiau and Te Wairoa and Ruakituri.* These people were descended from Te Tini-o-Maru, a still more ancient tribe, and which sprang from the god Maru. And Te Tini-o-Tauira occupied these lands many generations before the arrival of Mata-atua (canoe) or even of Horouta (canoe) from Hawaiki.

We do not know the history of that old, old race, but merely retain a few legends concerning them and their doings. But it was far back in the ages of darkness when Maahu and his people lived in this land, for he and others of his time were *atua* (gods) themselves, and held strange powers. There were other great chiefs also of Te Tauira who abode here. There was Manaia, who swam across the Sea of Wairau, and Tama, or Rongo-tama, from whom the ancient fort Nga Whatu-a-Tama is named, and Hau, and Rua, and Paka, with other semi-atua of the distant past.†

Now, Maahu must have lived many generations ago, for did he not engage in combat with Haere, the rainbow god, and each destroyed the other by supernatural powers?

LEGEND OF MAAHU, AND HAERE, THE GOD OF THE RAINBOW.

Tautu-porangi was the ancestor. He took Houmea-taumata and begat Haere-a-tautu, and Haere-waewae, and Haere-kohiko, and Hina-anga, and Hina-anga-tu-roa, and Hina-anga-whakaruru, and Moe-kahu. The three Haere became *atua piko*, or rainbow gods, and when the gleaming bow appears in the heavens we can distinguish which *atua* it is by the form and different colours. Among the Ngati-Kahu-ngunu Tribe, Kahu-kura is the rainbow god, and to the people below‡ it is Uenuku. I do not know if the Hina-anga sisterhood became *atua*, but Moe-kahu, the last-born of Houmea, was an *atua kuri Maori*, and appeared in the form of a dog. She is an *atua* of evil omen and destroys man. Moe-kahu is an *atua* of Nga-Potiki and Ngati-Kahu-ngunu. Maru-kopa-nui is another

---

* Rivers flowing into Hawke's Bay.

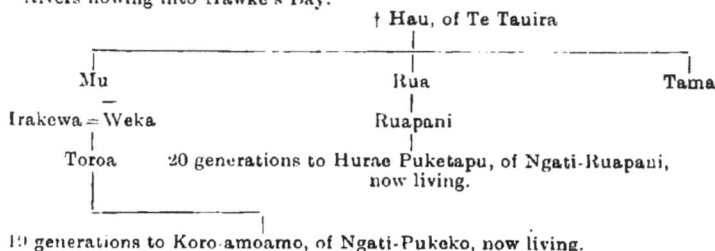

† Hau, of Te Tauira

```
                          Hau, of Te Tauira
                                 |
        -------------------------|---------------------------
        |                        |                          |
       Mu                       Rua                        Tama
        -                        |
  Irakewa = Weka               Ruapani
        |                        |
      Toroa       20 generations to Hurae Puketapu, of Ngati-Ruapani,
        |                            now living.
        |
        -------------------------|
                                 |
  19 generations to Koro-amoamo, of Ngati-Pukeko, now living.
```

‡ Below—*i.e.*, to the north.

atua ; he is represented by the glow seen above the horizon at eventide. Maru is a war god.

Tautu-porangi went forth to bear the *amonga,* or sacred food, to the god Wananga, whose *kauwaka* (medium of communication) was the priest Taewa. While engaged in this duty, Tautu was killed by Te Tini-o-kauae-taheke, an ancient tribe of very remote times. The word came to Houmea that Tautu-porangi was slain. Haere said, "Let us avenge the death of our parent." And Houmea replied, "Go forth to your duty, but be cautious, lest you cross the path of the gods—lest you tread upon the *aho,* which destroys man." So the children of Houmea went forth to attack their enemies, Te Tini-o-kauae-taheke; but on their way to the place of that people they trod upon the sacred *aho,* and perished by the *reti.* Their senses were destroyed by the *atua,* so fell they in the wilderness.

Haere and his brothers returned to their home. Houmea said, "What was the cause of your defeat?" The people replied, "We fell by the *reti* of our kindred, against whom we strove." Then Houmea gave them the means by which to overcome the sorceries of their enemies and retain life. She gave them the *taumata,* the *ahi,* and the *kete,* which are three very sacred and powerful incantations. Again they go out to attack the multitude of Kauae-taheke. On approaching the abode of their enemies they halted upon a hilltop and launched forth the sacred and powerful *karakia* known as the *ahi* or *kauahi* :—

Hika ra taku ahi Tu-e!	Kindle, then, my fire, O Tu!
Tu ki runga Tu-e!	Tu above, O Tu!
Tu hikitia mai Tu-e!	Tu, striding over, O Tu!
Kia kotahi te moenga Tu-e!	In one sleeping-place, O Tu!
Ko te taina, ko te tuakana Tu-e!	The younger and the elder brother, O Tu!
Kia homai Tu-e!	Give, O Tu!
Ki te umu Tu-e!	To the oven, O Tu!
Ki te matenga Tu-e!	To the death, O Tu!

This *karakia* being concluded, the *tohunga,* or priest, then uttered the following :—

Hika atu ra taku ahi Tu-ma-tere	I kindle my fire to Tu the swift,
Tonga tere ki te umu toko i a-i-i	Swiftly drag to the oven of wands,
Tere tonu nga rakau, tere tonu te umu-e.	Swift with the wands, swift to the oven.

And then—

Roki ai nga hau riri!	Be calmed the angry winds!
Roki ai nga hau niwha!	Be calmed the barbed winds!
Ka roki i nga rakau	Enervate the weapons,
Ka roki i nga toa	Make powerless the braves
Ka roki ki te umu-e	By the effect of the spell,
Ki te umu a Tu-mata-uenga——e!	By the spell of Tu the fierce-eyed! *

Having uttered these sacred spells, they then performed the *taumata,* which is a *karakia* to raise a great wind and cause an enemy to believe that no one will attack them on so boisterous a

* Tu is the god of war: he has many qualifying names indicative of his ferocity.

day. Also they used the Haruru or *kete* spell, the purpose of which was to draw the spirits of their enemies into a confined space and there render them powerless.

These great performances being over, then Hina-anga-whakaruru arose and exposed himself to the view of the multitude of Kauae-taheke, who cried, "It is a man!" Then Hina-anga bent downwards, and the multitude cried, "Not so; it is but a palm-tree. Behold! it is bent by the fierce gale." So Hina-anga kept deceiving those people, even until the shades of night fell.

In the dawning light the multitude of Kauae-taheke were attacked and defeated by the army of Haere and Houmea. And the *maawe*[*] of that battle was given to Moe-kahu, that she might bear it homewards. As she drew near this dog-woman barked loudly, and as she did so the knowledge came to Houmea-taumata that vengeance had been taken for the death of Tautu-porangi.

After this came the combat between Haere-a-tautu and Maahu. Each strove to destroy the other by means of the great powers they held, and both fell, each being slain by the other. The end of Haere was this : He was conveyed by Maahu to the *paepae*, where Noke, the earthworm, consumed him. As for Maahu, he was bewitched by Haere and caused to enter the sacred vessel Tipoko-o-rangi, in which he perished.

Such is one of the strange traditions handed down through many generations from the days of the ancient people, and which are known to but very few of the old men. Strange legends, many of these, localised here far back in the history of Aotearoa, but brought from older lands across the ocean in times long passed away. A reference to this ancient story is contained in a lament composed or adapted by Titi, of the Ngati-Kahu-ngunu Tribe, who was the medium of the god Po-tuatini, for his mother Rau-hanga and his relatives Pae and Puku, who had been slain by the Tuhoe people, some three generations ago :—

LAMENT OF TITI FOR RAU-HANGA.

Me he poko taku kuia Rau-hanga e ngaro nei,
Tenei te tamaiti wahine te tangi haere nei
Mauria atu ra kia taka i mua ki to korua nei aroaro
Ka toko ai e Toko-te-ahu-nuku.
Kotahi te kupu i hakiri ake i taku taringa,
Ko te mate o Tautu-porangi,
I haere ra te whanau ki te ngaki i te mate ;
Ka tu i te reti, ka ngawha te upoko,
Hoki ana ki te kainga.
I mauria i reira ko te kete, ko te ahi, ko te taumata,
Ko te ra kungia,
Ka mate i reira Tini-o-kauae-taheke.
Tena ko tenei, ina wai e ranga tena rakau tuki, rakau koki.
Rakau tu ki te tahataha
Ko Pito, ko Rere, ko Maika—e-e-a.

---

[*] A sacred emblem, often the head or war-lock of the first one slain.

Like a fire extinguished is Rau-hanga lost,
Whilst her little daughter ceaselessly weeps.
Take her then, and prepare her in your presence,
And let her be supported by Toko-te-ahu-nuku.
A story, imperfectly heard by my ears.
Relates the death of Tautu-porangi,
Whose family went to avenge his death.
They fell by the *reti*, their heads split open,
Causing them to return to their home.
Then were taken three powerful charms,
And a fourth to obscure the sun,
Which was the death of Te Tini-o-kaune-taheke.
As for my affliction, who shall avenge the sudden blow ?—
A blow delivered by the wayside,
Taking Pito, Rere, and Maika—e.

It seems doubtful if the genealogy of Maahu has been preserved by the Waikare people ; if so, no Pakeha has been permitted to acquire it. There were many ancestors of this name among various tribes of olden times, but none have hitherto been identified as the particular Maahu who bestowed upon mankind the boon of causing the Waikare-moana Lake to be formed. Now, however, the Kaumatua comes to our rescue with a *waiata*, or song, in which the full name of Maahu appears, and this will set conjecture at rest on this point.

LAMENT.

Lament for Te Whenua-riri, a chief of the Ngati-Kahu-ngunu Tribe, killed by the Ngapuhi Tribe at the fall of Titirangi Pa, near Waikare-moana, about the year 1824 :—

I tawhiti ano te rongo o te pu,
I ki ano koutou. Ma wai ra e homai?"
Ki te kainga o Maahu-tapoa-nui,
Ki a Tu-ma-tere ra,
Ki te oke ki te pae.

Whilst distant was the fame of the guns,
All said, " Who will dare to bring them here ?"
To the home of Maahu-tapoa-nui,
To Tu the swift, indeed,
To strive within our bounds.

E koro ! ki nui, ki patu, ki tata-e !
I te rangi maori
He mea ra kia kapi te waha
Ka kitea a rikiriki,
Ka peke mai Tini-o-Irawaru,
Hei poke mo koutou.

O Sir ! of great, of warlike words and blows,
Heard in this ordinary world,
'Twas so said that mouths should bo closed.
Now, indeed, is seen inconsolable grief,
Spring forth the descendants of Irawaru*
To worry and tease you all.

Takoto mai ra E koro E !
Koutou ko whakahina
I te hara kohuru
Nau era ngohi,
E ware koutou ki Te Toroa ma ?
Tera te Poturu nana i kai atu.
Takoto mai ra E koro E !
I roto o Tauri
I hea koia koe ka aho ai i to tapuwae
Ata tu mai ! ata tu hihiko mai,
He hihiko hei hiki mai i a koe
Ki te rangi i runga ra
Ko aua wai ano to mata nei whakataha.

Rest thee there, O Sir !
Thee and thy grandchildren,
Through the evils of murderous war.
Thine are the slain.
Have all forgotten Te Toroa's death ?
Still lives Poturu, who consumed them.
Prone thou liest, O Sir !
In the vale of Tauri
Where wast thou that thou charmed not thy steps ?
Stand forth ! arise with vigorous strides,
Strides that will bear thee on
To the heavens above us ;
To those other waters turn thy face aside.

* Irawaru, the father of dogs.

Toi = Mokotea.

Iho. Whaitiripapa.
Te Marangaranga.
Te Uira-i-waho.
Tu-rere-ao.
Mai-ki-te-kura.
Te Rupetu.
Hatonga.
Maahu-nuku.
Maahu-rangi.
Maahu-tapoa-nui.
Te Rangi-taupiri.
Tamaka.
Te Ao-whakahaha.
Kuia-rangi.
Tihore = Kopura-kai-whiti.

Pau-mapuku.
Hine-tara.
Tama-ka-uru.
Te Wherutu.
Tu-whakarau.
Te Whiua.
Te Rangi-pakakina.
Kurukuru.
Tikitu I.
Tikitu II.
H. Tikitu.
Te Uri Kore.
Titirahi.

Now, here we have the full name of friend Maahu given, and Maahu-tapoa-nui is known to genealogists as an ancestor of the Ngati-Awa Tribe, of the Bay of Plenty district, on the aboriginal side—that is, he was of the people found dwelling here when the ancestors of the present Maori migrated to New Zealand. Different lines give from eighteen to twenty-one generations from Maahu-tapoa-nui down to the present time. It is therefore probable that he flourished in Tuhoe land about five hundred years ago. His position is shown in the genealogical table printed in the margin.

But the great work performed in the time of Maahu was the formation of the Waikare-moana Lake by his son Hau-mapuhia, and we will now give the generally-received version of that most ancient and wondrous legend.

The Legend of Hau-mapuhia.

It is the story of Hau-mapuhia, and how this sea of waters was formed from the dry land in the days of yore, also the explanation of the name of this great lake.

Maahu took Kau-ariki to wife; their child was Hau-mapuhia. They dwelt at Wai-kotikoti, at Wairau-moana, and Hau', being well cared for, grew to be a fine young man, though some say that Hau' was a girl. And it happened, as the shades of evening fell, that Maahu bade Hau' go to the spring called Te Puna-a-taupara and bring thence a gourd of water. But Hau' was unruly, and refused to go to that spring, at which Maahu was greatly enraged. So he took the gourd and proceeded to the water himself, where he stayed so long that Hau' went after him. On his arrival at Te Puna the thought came to Maahu that he would kill his child for being disobedient; and he took Hau' and thrust him into the water and held him below the surface thereof. Then Hau-mapuhia called on the gods of the ancient people, and they came to his aid. This they did by endowing him with great and wondrous powers such as demons possess. Hau-mapuhia, son of Maahu, was transformed into a *taniwha*—a water god. Armed with these strange powers, Hau' forced his way through the solid ground and formed the great hollow in which lie the waters of Waikare. Previous to that time it was all dry land. Such a *taniwha* is called by us a *tuoro* or *hore*. And it was in forcing his way through the ground seeking an outlet that Hau-mapuhia formed the many arms and inlets which you see around this lake. The fierce struggle by which he forced his passage from Te Puna-a-taupara, which is the *tino* of Waikare-moana, so agitated the waters which followed him that the lake has ever since been known as the "Sea of the Dashing Waters." (*Ka hokari nga ringa me nga waewae, katahi ka pokare te wai, koia i kiia tona ingoa ko Wai-kare moana—ko te pokaretanga o te wai.*)

The first attempt made by Hau-mapuhia to escape was towards the west, that was how the Whanganui inlet was formed, even to Herehere-taua, where he was stopped by the great bulk of Huia-rau. He then turned and tried another direction, thus forming the Whanganui-o-parua inlet. But the great ranges again held him, and, after forming the other bays of Waikare, Hau the demon turned to the east, whence he heard the roar of the Great Ocean of Kiwa[†] in the far distance, and the thought came that it would be well to reach that great ocean before the light of day appeared. So Hau again forced his way downwards at Te Wha-ngaromanga and strove to burrow through the ranges to the Moana-nui-a-Kiwa (the ocean). But when he reached the *komore*, where the waters rush forth, he became fixed there, and so lies even to this day. Maybe the emerging into the light of day caused the power of Hau-mapuhia to fail, or maybe the gods were alarmed lest his great efforts should release the waters of the newly-formed sea of Waikare.

And as Hau-mapuhia lay there in that ravine he moaned aloud in wailing tones, and Maahu, who had gone to the great ocean, overcome with remorse at having slain his son— Maahu heard his off-spring wailing afar off, and he called upon the koiro and the tuna, the kokopu, maehe, and korokoro, and other fishes to go by the River Waikare-taheke, which reaches the great ocean, and ascend to where Hau-mapuhia lay, that they might serve as food for his child. But the koiro (conger-eel) would not face the fresh water, and the tuna (fresh-water eel) could not pass the Waiau River, and the maehe and korokoro (lamprey) were the only two fish which reached the Waikare-taheke River to serve as food for Hau-mapuhia, and it is said that the korokoro is not found in any other stream in the district.

And Hau-mapuhia still lies there where he emerged, transformed into stone. His head is down hill and his legs extend up the hill-side, and the lake-waters, rushing forth from the hill, pass through his body to form below the Waikare-taheke River, as you shall see. Also you may see his hair floating and waving in the foaming waters : this hair is in the form of what we call kohuwai (a water plant).

Then Maahu left these lands, and went far away to Pu-tauaki,

---

* *Tino*, the precise spot from which a district, &c., takes its name.

† Te Moana-nui-a-Kiwa, the Great Sea of Kiwa, a term applied to the Pacific Ocean by the Maoris.

where he remained ; but his heart was dark for his deserted home at Wairau-moana and the lands of his ancestors, which he had deserted. Even so he raised his voice and lamented,—

Kaore hoki i te roimata te pehia kei aku kamo	Alas, the tears weigh heavy in my eyes,
Me he wai utuutu ki te Wha-ngaro-manga—e	Like water gushing forth at Te Wha-ngaromanga,
Ko Hau-mapuhia e ngunguru i raro ra—e-a.	Where Hau-mapuhia rumbles down below.

Such is the legend of Hau-mapuhia and the formation of Wai-kare-moana. A strange legend and an ancient, viewed from the standpoint of an unlettered people possessing no knowledge of the graphic art, and relying entirely upon oral tradition. It originated probably in the widespread and universal desire implanted in the human mind to assign a cause and origin to all material objects and manifestations of Nature.*

There is another class of legend which obtains in several districts, the names being altered to fit local circumstances. Of such is—

THE STORY OF RAKAHANGA : A LEGEND OF THE TAUIRA, OR ABORIGINAL PEOPLE OF WAIKARE-MOANA.

Rakahanga-i-te-rangi was a *puhi* (a virgin, a betrothed girl) of ancient times, and dwelt with her people at Turanga (Poverty Bay) The fame of her beauty reached the chiefs of the multitude of Tauira, who dwelt by the shores of the " Rippling Sea of Waikare." So the thought grew, " Let us acquire this famous and lovely lady as a wife for one of us." Then Rongo-tamá, a descendant of Whaitiri, with Hau and Rongo-i-te-karangi, formed with great care a party of seventy men, who were so selected as to be all men of fine appearance and of equal size. They were also well trained in the various kinds of songs and dances known to the ancient people. Their object was to show what fine men the Tauira were, and how greatly accomplished, that Lady Rakahanga might choose a husband from the visiting chiefs. This kind of expedition is termed a " *kai tamahine*."

When about to commence their journey, the council of the chiefs decided that Hau was too ugly to form one of the party, as he wore a beard, so it was decided that he should be left behind, for it would never do to parade such a plain fellow before the famous beauty Rakahanga. Thus the party left without him ; but Hau, who was a man endowed with certain wondrous powers, hid himself beneath the *punake*, or bow, of the canoe which conveyed the party to Turanga, and so arrived safely at that place, where the men of

---

* There is little doubt that the lake was formed by a vast land slip, now covered with forest, which fell from the slopes of the mountains on the east of the outlet, and filled up what was formerly a valley. Probably this took place before the advent of the Maori ; but he is quite equal to understanding the cause, and, with his love of the marvellous, to inventing a supernatural reason for it — EDITOR.

Te Tauira landed and proceeded to the village where Rakahanga lived. When they were out of sight, Hau came forth from his place of concealment and hid himself until nightfall.

In the evening all met in the *whare-tapere* or amusement-house of the pa, where the visitors were to show their various accomplishments, in order that one of their number might find favour in the eyes of Rakahanga. And some of the village people were sent to collect fuel, which was to be carefully selected in order that the fires in the *whare-tapere* might burn clearly and not smoke. But that deceitful Hau drew near, and caused those fires to smoke dreadfully, by means of the following *karakia* :—

> Ka pu auahi ra runga,   Gather together the smoke from above,
> Ka pu auahi ra raro.   Gather together the smoke from below.

This incantation, in fact, produced so much smoke that many of the people, including our heroine, rushed forth from the house into the darkness of night, where Rakahanga was met by Hau, who, favoured by the gloom of night, which concealed his ugliness, and more so by a spell termed *tau-patiti*,* managed to ingratiate himself with the much-sought Rakahanga. So much indeed did he take that lady's fancy that she determined to choose this man as her husband, and so she marked him by pinching his forehead, that she might know him when they met in the light of day.

The next morning when all were assembled, Rakahanga proceeded to search for the man she had chosen as her husband, looking closely at each of the visitors in order to detect her mark. For a long time she failed to detect this marked man, until at last Hau appeared, and then poor Raka' was much disgusted to find in him so ill-favoured a man. And the other chiefs were much disgusted that this ugly fellow should win so charming a girl. So the chief Kiwi, disregarding Hau's claims, took Rakahanga as a wife for himself, and, accompanied by his friend Weka, they started by an inland track to Waikare. The deserted Hau was indignant at losing his promised wife, and started in pursuit. As he journeyed on through the forest he came to where two men named Tane-here-ti and Tane-here-pi were spearing pigeons in a tree. Hau inquired of these men whether they had seen any travellers pass by. They replied, " Yes, two men and a woman have passed here." So Hau went on until he reached Waimaha, where he overtook Kiwi and his companions. Hau, armed with his famous greenstone battle-axe, Hawea-te-ma-rama, at once attacked the two men, and slew Kiwi, but Weka and Rakahanga escaped and fled far away into the forest. Friend, lest you be misled, this is not the same Weka who married Toroa, for that was long after.

And so Weka and Raka' travelled on across the great hills until they came to Te Reinga Falls, on the Wairoa River. The night was falling when they arrived at that place ; the darkness settled down as they traversed the summit of the dread cliff above the falls ; a

---

* *Tau-patiti.* This *karakia* is now called by us an *iri*.

3

great fear came upon Raka', and she said, "Friend! let us be cautious, lest we fall from this great cliff." But Weka said, "Fear not; this is the track by which we go." But as he spoke they came to the highest and most dangerous part of the cliff, and Weka quickly turned and thrust the poor girl over the edge of the rock into the black chasm below. The reason of this act was the fact that Raka' had favoured the ugly man Hau at Turanga-nui-a-Rua.

Thus, in that fearful chasm, perished Rakahanga-i-te-rangi, the famous *puhi* of old. And her last words were, "*He po Rakahanga i raru ai*" ("By darkness was Rakahanga confounded").

Those who have heard the story of Wairaka of Mata-atua canoe will at once recognise the resemblance that this story of Rakahanga bears to it. It may be noted that many of the stories told concerning the ancestors of the present Maori people are also related by the descendants of the pre-Maori aborigines as having occurred in the times of their ancestors, and long before the arrival of the historic Maori fleet from the Hawaikian fatherland in about 1350. Were the origin of these old legends known, it is probable that the ancient people would carry the day as the originators thereof.

Another tradition of this kind is—

The Legend of Rua and Tangaroa.—From the Ancient Nga-Potiki Tribe.

[Rua was a famous ancestor, and lived in very remote times. He was the inventor of carving, hence the expression, "*Nga mahi whakairo, nga mahi a Rua.*"[*] There are many terms and place-names in connection with this ancestor. Te Whatu-turei-a-Rua is an ancient term for the meal made from the berries of the hinau-tree; Turanga-nui-a-Rua is the name of Poverty Bay; Te Whakaki-nui-a-Rua is a lagoon at Te Wairoa; Tamaki-nui-a-Rua is the name of the Seventy-mile Bush; Te Awa-nui-a-Rua was the ancient name of the Whanganui River, but it is not clear whether such was the name given by the aborigines of the latter river, or whether it was so termed by Nga-paerangi of Horouta canoe, which people held the valley of the Whanganui some generations before the Aotea migrants arrived there.

The following tradition of Rua and Tangaroa is a most singular and interesting remnant of an ancient mythological system, though unfortunately it is impossible to obtain an explanation of it at this late day. Tangaroa is the Polynesian Neptune, the tutelary genius of the great ocean of Kiwa; indeed, in many of the islands of the Pacific he is the Supreme God and Creator. He is represented in this legend as dwelling in a house beneath the ocean, and his tribe is composed of the fishes of the great sea. Maroro, the one member of Tangaroa's tribe who escaped, is the flying-fish. As in other old-time legends of these gods and their subjects, the characters are all endowed with the power of speech.

---

[*] "The art of carving, the art of Rua."

The name of Rua is very common in the ancient traditions of Waikare-moana and the aboriginal Nga-Potiki of Maunga-pohatu, but it does not appear to be now known which Rua is alluded to.

Rua dwelt in his place in the days of yore, in the very distant times, remote beyond expression. The thought came to him that it would be well to visit Tangaroa-o-whatu. So Rua went to the house of Tangaroa, and on his arrival found that being jubilant over the fine appearance of his house, which, he asserted, had been embellished with wondrous carved figures by Hura-waikato. And Tangaroa said to Rua, "Do you come with me and behold my fine house, for doubtless you came to admire the grand work of Hura-waikato." Now, when Rua saw the house of Tangaroa he was much astonished to find that the wondrous carving of Hura was no carving at all, but simply painted figures, such as are seen on the rafters of our houses. Then Rua asked, "Is this your famous carving?" Tangaroa replied, "Yes, this is the carving." Rua said, "Do you come to my place and see what real carving is," for Rua was the father of the art of carving, and hence comes the expression "*Nga mahi whakairo, nga mahi a Rua*" (see *ante*). And the house of Rua was a truly brave sight, so adorned was it with carving and so fine were the figures.

On a certain day Tangaroa set forth to visit the dwelling of Rua. As he approached the house, and while some little distance off, he observed the carved human figure (*tekoteko*) which adorned the front of the house. So he greeted this figure with the words "*Tena ra koe*" (Salutations to you), and then, walking up to the *tekoteko*, he proceeded to embrace it, or *hongi* (rub noses), according to our ancient fashion, not thinking but what this beautiful figure was a living man, so fine was the carving of Rua. As Tangaroa entered the house Rua laughed at him, saying, "This is indeed carving, you see how you have been deceived by it." Then was Tangaroa overcome with shame. He therefore returned sad-hearted to his own place, but before he did so he managed to obtain the pet *koko* bird (tui) of Rua, which was a clever bird, and much prized by its owner. This evil act he committed under the cover of darkness, and then carried the bird off to his own place, which lies within the ocean. When Rua discovered the loss of his *koko* he was much grieved, and at once sallied forth in search thereof. After wandering afar off he came to the shores of the ocean, and as the tide broke and flowed back to the Waha-o-Te-Parata* Rua heard the missing *koko* singing within the great ocean—that is, at the home of Tangaroa. So Rua resolved to obtain his pet bird, and therefore entered the realm of Tangaroa. On his arrival at the latter's house he found that Tangaroa was absent, having gone forth into his great domain. The only beings remaining at the *kainga* were Tatau the doorkeeper and the *koko*. Then Rua asked of Tatau, "Where is Tangaroa"? The doorkeeper

---

* Te Waha-o-Te-Parata : The Maoris account for the tides of the ocean by saying that a huge monster dwells at the bottom of the sea named Te Parata, and that it is the inhaling or exhaling of his breath that causes the tides. EDITOR.

replied, "He is abroad in the ocean seeking and slaying food." Rua said, "Will he return to this place?" "When the shades of evening fall he will return," said Tatau. Then Rua instructed Tatau how to act when Tangaroa returned. He said, "When the day dawns and Tangaroa cries out to you, ' Tatau E! draw aside the door,' do you repeat these words,—

> E moe. Ko te po nui, ko te po roa,
> Ko te po ka whakaua ai te moe
> E moe !"

And then, when the rays of the sun come steeply down, do you draw aside the door of the house, that the sun may shine with strength into the home of Tangaroa."

Tangaroa returned home in the evening and entered his house, where he and his tribe slept. When morning came and he thought that daylight must be at hand, he cried, " O Tatau! draw aside the door." Then Tatau repeated this incantation :—

> Sleep on ! Through the great night, the long night,
> The night devoted to sound sleep,
> Sleep on !

So Tangaroa again slept. When the sun waxed strong, then the sliding-door was opened by Tatau. The sun flashed into the abode of Tangaroa and destroyed him and his people. Maroro was the lone survivor. *Heoi !*

Of a more singular nature still is—

THE STORY OF RUA-KAPANA : A LEGEND OF THE ANCIENT PEOPLE, As preserved by the Nga-Potiki people of Maunga-pohatu, and told by the Kaumatua.

Pou-ranga-hua was a chief of the ancient people of the land. He took to wife Kanioro, who it is said was a sister of Taukata, who brought the knowledge of the *kumara* to the aborigines, to Toi of old. Pou-ranga-hua's place of abode was at Turanga (Poverty Bay), and his were works of wonder in the days of old. One of these labours was the formation of a lake at Te Papuni, which he effected by means of a *karakia* or incantation, which spell contracted the hills and made them close in across the valley. In my young days I thought that it would be a good idea to drain this lake and so obtain a vast quantity of eels which frequented it. So I took with me a hundred of my young men of Nga-Potiki, and we commenced to dig a large ditch from the lower side of the lake. And as we neared the lake the great body of waters broke in upon us, and we fled swiftly, being nearly overwhelmed by the great rush of the flood. So great indeed was its force that the waters broke out two more small lakes which lay below, and we lost the greater number of those eels.

---

* This *karakia* is termed a *rotu*, its effect being to cause people to fall into a deep sleep.

And Pou' bethought him of building a house at Turanga-nui-a-Rua. When the house was finished he set forth on a journey to the Kauae-o-Muriranga-whenua in order to obtain *takuahi* (stones) for a fireplace. When out upon the ocean in his canoe, the wind known as Te Hau-o-pohokura arose and drove his canoe far away across the dark waters. It is said by some that the dread *taniwha* (demon) Rua-mano conveyed Pou' over the Ocean of Kiwa, and he was cast ashore at Pari-nui-te-ra, at Hawaiki. When Pou' looked about him he saw that he was in a strange land, for which reason he was sore dismayed. Then he came to the people of the land, and among them was the great chief Tane-nui-a-rangi, who took Pou', the castaway, to his own place, and treated him with much kindness. And Pou' dwelt among the kindly people of Hawaiki, for the land was a fair land and a bountiful.

During all this time his wife Kanioro remained within the *whare-potae* (house of mourning) in this land of Aotea-roa, and there was no peace for her; neither did the bright sun shine, for she mourned the death of her husband, of Pou-ranga-hua.

And as Pou' dwelt in that strange land the thought grew, that he must return to the White World of Maui of old,[*] that he might greet his wife Kanioro of Nga Tai-a-kupe. Then he said to Tane, "How may I return to my home, to Aotea-roa?" Tane said, "Get your ancestor Tawhaitari to take you across the great waters." Now this Tawhaitari was a huge bird which belonged to that strange land. So Pou' obtained the services of Tawhaitari to bear him back to Kanioro; but first he went to the summit of Pari-nui-te-ra and obtained there two baskets of *kumara* (sweet potatoes), for that valued food was then unknown by our ancestors here. The name of one basket was Hou-takere-nuku, and of the other Hou-takere-rangi. Then he obtained the two *kaheru*, Manini-tua and Manini-aro. All these he secured upon the back of the bird, and then mounted himself. The great bird then attempted to rise and commence its long flight, but could not rise on account of the heavy burden. So Tawhaitari was rejected by Pou'." Then Tane said, "Fetch your ancestor Te Manu-nui-a-Rua-kapana—the great bird of Rua-kapana. Pou' did so, and placed the burden on the back of Rua-kapana. Tane then spoke, "Farewell! Go forth to your home which lies far away across the dark waters; and do you keep firmly to my words—be kind to your ancestor, to Rua-kapana. Do not allow it to land in your country, but when, on nearing the shore, the bird shakes itself, do you quickly alight, that the bird may return safely here."

So the great bird of Rua-kapana rose into the air, and stretched out across the Moana-nui-a-Kiwa, bearing Pou-ranga-hua and his prize to the home of Kanioro. Now, the object of Tane in his warning to Pou' was this: On the summit of the Mountain of Hiku-rangi,[†] which lies far away towards the rising sun, there dwelt a *tipua*,

---

[*] Aotea-roa, or New Zealand.
[†] Hikurangi, a high mountain near the East Cape, New Zealand.

or demon, in the form of an old man, whose name was Tama-i-waho, An *atua* (god) was this Tama, possessed of evil powers. So great indeed was his command of sorcery and evil arts that no living thing could pass that dread mountain, all were destroyed and devoured by Tama, the goblin of Hikurangi. There was one time only during which this evil place might be passed, and that was when the sun declined so far as to cast its rays into the face of Tama, which so dazzled his eyes that he was unable to see. That was the only salvation for man—the fact that Tama could not see during strong sunlight.

Thus came Pou-ranga-hua and Rua-kapana from far Hawaiki. As they approached Hikurangi, they waited until the sunlight slanted into the eyes of Tama, then they fled quickly past that dreadful spot ; and as they did so Tama-i-waho cried, " Who is this ascending the mountain of Tama-nui-a-rangi ? " But when his sight came back to him Pou' and his bird friend had passed by. As they approached the shore at Turanga, the bird shook itself, as a sign to Pou' that he should descend and leave Rua-kapana to return safely to Hawaiki. But Pou' refused to get down, and kept his seat on the back of the bird, compelling it to take him to his home at Turanga. And the great bird of Rua-kapana knew then of the doom which awaited it should it pass within the evil shadow of Hikurangi. Then Rua-kapana said, " O Pou' ! what an evil man art thou." But Pou' only said, " Pou' returns but once, the door is closed on the road to Hawaiki." Such were the words of Pou-ranga-hua.

And as they approached Turanga, Pou' reached under the wings of the great bird and plucked therefrom the fine plumes, which he threw into the sea. And from these plumes cast into the ocean at that place there grew a kahika,* the name of which is Makauri, and that tree still bears fruit out in the ocean. And a branch of that kahika was broken off and cast inland. From that branch came the fine forest which stands between Ma-karaka and Te Waerenga-a-hika, which forest is also known as Makauri.

So Pou' compelled Rua-kapana to bear him to land, even to the mainland, and the great bird set forth to return to Hawaiki, but on passing Hikurangi it came within the influence of Tama-i-waho, the ogre of the mountain, and was destroyed by that monster. So perished the Manu-nui-a-Rua-kapana.

Pou-ranga-hua planted his seed *kumara* in the cultivation at Manawa-ru, at Turanga. That is how the *kumara* was brought to that district.

He then went to his home, to the place where he left his wife. On his arrival he found the house shut up and bearing a deserted appearance, being overgrown with mawhai.† Within the desolate house was Kanioro, mourning for her husband.

---

* Podocarpus dacrydioides.
† Mawhai, *Sicyos angulatus*, a plant.

Then Pou' tapped the door of the house, and Kanioro cried, "Who is that tapping outside?"

"It is I, O Kanio! It is Ranga-hua."

The voice of the woman was heard:

"Ranga-hua was swept away by the Hau-o-pohokura."

"Give to me some of thy valuable treasures, O Kanio!"

"For what purpose?"

"As a reward, O Kanio! A reward for the Manu-nui-a-Rua-kapana."

Then Kanioro pulled aside the door, and Ranga-hua entered their house and kindled a fire therein; and Kanioro gave her treasures unto Pou-ranga-hua, even as he had demanded, for her joy was great. But from that time she discarded the name of Kanioro, and took that of Tangi-kura-i-te-rangi.

And it is said that the Ngati-manu-nui Hapu of Tuhoe, who reside at Te Umu-roa, derive their name from the Manu-nui-a-Rua-kapana.

Meanwhile, Tane-nui-a-rangi was anxiously awaiting the return of the great bird of Rua-kapana to Pari-nui-te-ra, for the time he had arranged for it to arrive had passed. Then the knowledge came to him that the bird had been slain by Tama, the "Ogre of the Enchanted Mountain." So he summoned Taukata, and told him to set forth in search of the lost bird and the person who had killed it. And Tane said to Taukata, "By this sign shall you know the slayer of Rua-kapana—that is, by the sign of the *niho-tapiri*."* These things were quite clear to Tane on account of his great powers in magic.

And Taukata came from far Hawaiki to this land of Aotea-roa, being conveyed over the vast ocean by a water-demon, such as were plentiful in the days of our ancestors, so wondrous were the works of old.

The blackness of night was descending upon the earth when Taukata came to the "Enchanted Mountain"—to Hikurangi. He then concealed himself near unto the doorway of the house of Tama-i-waho, where he busied himself in uttering the most potent incantations—the most sacred spells. Then he entered boldly the abode of Tama, the dread ogre, and seated himself among Tama's people. And Taukata listened to the talk going on around him, but could not understand it, as it was all nonsense and mere gibberish. When it came to his turn he spoke these words: "*E kore e tangi te whati-tiri, no-e, no-e*" (the thunder will not sound, *no-e, no-e*). At the same time he patted the shoulder of the man whom he suspected of being the slayer of Manu-nui. This so amused the assembled people that they all laughed, showing their teeth as they did so. Here was Taukata's opportunity. He gazed intently. Aha! the *niho-tapiri* was seen. Then he cried, "Let us extinguish the fire and all go to sleep." It was done; the people slept. Behold! it is Tau-kata who produces a *kete*, a large basket, into which he places the

* *Niho-tapiri*, uneven teeth, growing in an irregular manner.

body of the owner of the *niho-tapiri*. Then the *karakia*, the magic spell, to induce profound slumber—oblivion. It was the *rotu*:—

E moe, E moe! Ko te po nui, ko te po roa	Sleep on! sleep on, the great night, the long night
Ko te po i whakaau ai te moe. E moe!	The night devoted to sleep—Sleep on!

Such is the *rotu* spell to cause men to sleep. The *rotu-moana* is a different *karakia ;* it is to cause the ocean to sleep or become calm, that canoes may pass over it in safety, and the other term for it is *awa-moana.*

So Taukata secured the sleeping man in the great *kete*. So sound was the slumber of that man that he never awoke through all the long journey to Hawaiki.

It is Pari-nui-te-ra, the land of plenty. The light of day comes, the sun shines brightly, and Taukata has returned from the mountain of Tama the ogre—returned with the slayer of the great bird of Rua-kapana. It is the day of vengeance. The multitude of the land are assembled ; the great chief approaches ; it is Tane-nui-a-rangi—Great Tane of the Heavens. Taukata stands by the body of his still sleeping prisoner. He speaks : " Awake ! lest you think that you sleep in thine own place." Then the man awoke—awoke and looked forth upon the land ; saw the strange land—land of the Great Cliff of the Shining Sun ; saw the multitude assembled ; saw Taukata and Tane : then the thought came, it is death.

So they killed that evil man, and ate him, as he had eaten their *tupuna*, Te Manu-nui-a-rua Kapana. *Heoi !*

The above is a most singular legend, and interesting from many points of view. In the first place, it appears to be a local adaptation of the Polynesian tradition of Tinirau and his pet whale Tutu-nui, which was slain by Kae ; in fact, the stories are almost identically the same. The hero, Pou-ranga-hua, was a chief of the ancient people of New Zealand, and is well known to their descendants in Tuhoe land. Te Hau-o-pohokura is a sea-wind which blows in the spring of the year. Rua-mano was a *taniwha* or demon of olden times, who is said to have resided at Te Papuni during his latter days. He is said to have been the offspring of Tutara-kauika, which last appears to have been a kind of emblematical term for the whale.

The great bird of Rua-kapana is a decided puzzle, but it is possible that we may yet be enlightened as to what it was ; for Nga-Paerangi, a tribe of Whanganui, have retained a legend anent one Rua-kapanga and a huge bird of olden times. Now, this tribe is descended from Paerangi, son of Paoa, who came from Hawaiki in the canoe Horouta, and landed on the East Coast near unto Hiku-rangi. It is possible that they have preserved this legend, and that the names mentioned therein have become somewhat altered during the lapse of many generations.

The nature of the reward or payment given by Kanioro to Rua-kapana is not clear, but the light may yet shine thereon. There is some old, half-forgotten story of Kanioro as having been an *atua-pounamu*, or guardian of the greenstone, most prized of Maori treasures. It may be that the *kumara* was given to Pou-ranga-hua, and the services of Rua-kapana loaned to him, on condition of his sending back the precious greenstone in payment thereof. The two *kakeru* given to Pou' at Hawaiki are said to have been chaplets or head-dresses.

The name of the human ogre of Hikurangi—Tama-i-waho —is also that of the *atua* who visited Te Kura-nui-a-monoa, wife of Toi, and by whom she had Oho-mai-rangi, also known as Oho-matua-rau ;—though some tribes claim that Puhao-rangi takes Tama-i-waho's place. There is some mention in the Rev. R. Taylor's " Te Ika-a-Maui " of a legend concerning a great bird which existed on Hikurangi in olden times. The Tuhoe have a tradition of a bird called hakoke which frequented cliffs and mountains, but which has been long extinct. In fact, the legend appears to be a confused and half-lost fragment of a very ancient folk-lore system.

The *whare-potae*, or *whare-taua*, is, literally, a mourning-house. If a man of distinction dies, his son or near relatives remain for some time in the *whare-potae*, never venturing forth, and only taking food during the night-time. After a certain lapse of time a human sacrifice is made, to take the *tapu* off these imprisoned mourners : *hei heuenga mo te whare-potae*, or dispersal of the mourners. When Taupoki died, at Wai-koti-koti, a slave was sacrificed for this purpose, the body being cooked on the river-bank where the camp of the soldiers stands.

We will now cease these old tales and speak of the wars which waged between the tribes of Tuhoe land and Ngati-Ruapani, of Wai-kare-moana.

WARS OF NGATI-RUAPANI AND TUHOE.

The Ngati-Ruapani Tribe of Waikare-moana are an ancient people, and have dwelt in this district for many generations. The principal hapus or sub-tribes are Ngati-Hine-kura, Ngati-Tahu, Ngati-Haua, Ngati-Mate-wai, and Ngai-Te-Amohanga. Among them are also a few Ngati-Ira, who came from Opotiki to the lake, *via* Rua-tahuna, about four generations back.

The hapus of the Tuhoe Tribe are : Ngati-Hine-kura or Ngai-Te-Riu (the sub-hapus being Ngati-hora-aruhe, Ngai-te-ua, Ngai-te-rurehe, and Ngati-rohe) ; Ngai-Tawhaki (the sub-hapus being Ngati-Tamakere, Ngati-Koro, Ngati-Tuhea or Ngati-Tu-haere-ao, and Ngati-Taokaki) ; Ngati-kaira ; Ngai-tumatawhero ; Ngati-rere-kahika ; Nga-Potiki ; Ngati-Ha ; Ngati-Maru ; Te Upoko-rehe ; Patu-heuheu (allied) ; Tuhoe-potiki ; Ngai-taraparo ; Ngati-te-umu-iti ; Ngati-Kakahu-tapiki ; Ngati-Ruri ; Ngati-Hamua ; Ngati-Koura ; Ngati-Rongokarae ; Ngai-Turanga ; Nga-maihe ; Ngai-Tama ; Ngati-manunui ; Tama-kai-moana, or Ngati-huri, being the descendants of the ancient Nga-Potiki.

In the time of Hatiti, cf Nga-Potiki — that is, some twelve

generations back—fighting commenced between Tuhoe and Ngati-Ruapani, of Waikare-moana. A party of the latter tribe, under Taranga-a-kahutai, crossed the Huia-rau Range and attacked the Nga-Potiki pa of Raehore, which was situated on the range above the Rua-tahuna Stream, taking the pa and killing Hatiti, son of Potiki the Second. Tuhoe collected their men from many isolated settlements and drove the Ruapani back across Huia-rau.

Nga-Potiki then raised a *taua* (war party), under Tahaki-a-nina and others, and crossed the mountains to Waikare, where they found their enemies at the mouth of the Opu-ruahine Stream, at the Whanganui Inlet. The two tribes met and fought at Te Ana-putaputa, beneath the steep range on the eastern side of the small branch of the Whanganui arm of the lake. On this narrow strip of beach the battle waged fiercely for some time, with the result that Ruapani were defeated, losing the chiefs Taua-tu and Taunga-atua, with many men of lesser rank. As their enemies fled along the base of the cliff, Tuhoe pursued them, killing numbers among the rocky boulders which line the lake-side.

Ngati-Ruapani in their turn now marched on Tuhoe, at O-hauate-rangi, on the ranges near Rua-tahuna. At O-te-rangi-o-raro they captured the wife of Hapopo, who, however, contrived to escape, and fled quickly to her husband, whom she apprised of the oncoming *taua*. Hapopo at once set about consulting his *atua* (deity) like a true Maori, in order to ascertain the truthfulness or otherwise of the story, and the possible result to himself of an appeal to arms. This *atua*, Tu-a-kahu-rakiraki, which is an *atua whakaepaepa*, appeared to treat the matter very lightly, merely repeating the word " *Tikore ! Tikore ! Tikore !* " thus conveying the meaning that no danger existed. This set the mind of Hapopo at rest ; but his sublime faith in the god was ill repaid, inasmuch as he was shortly afterwards slain by the war-party. The following well-known and much-plagiarised saying, " *Na Tu-a-kahu-rakiraki, waiho te mate ki a Hapopo,*"[*] is applied to the false prophecy of this *atua*. Tuhoe now gathered their available warriors in the vicinity and attacked the Ruapani, who were defeated and forced to retreat homewards, having lost the chiefs Taranga-a-kahutai, Whatai, Te Kawakawa, and Te Tuhinga.

To square accounts with the Tuhoe for the death of the above chiefs, the lake-men then mustered a force which marched by Opuruahine across the ranges to Maunga-pohatu, where they attacked Nga-Potiki of that secluded district, but were obliged to fall back, after losing Tai-ka-ea, Haua, and Te Neinei, leading men of the party. Peace was then established between the two tribes, and continued for many years.

Then, again, about four generations ago, war arose in the land. Te Umu-ariki and Koko-tangi-ao, of Tuhoe, were killed by Ngati-Ruapani at Whanganui. The body of the latter was mutilated by

---

* " 'Twas Tu-a-kahu-rakiraki that abandoned Hapopo to death."

his slayers, who made merry over it with bitter jests. Word came to Rua-tahuna of this dire insult, brought by refugees from the wrath of Waikare. Worse still, it was known that Mokoa of Ruapani had vowed that he would use the body of the Tuhoe chief, Tipihau, as bait for his *hinaki* (eel-basket). This was an insult of an appalling nature, and by all rules of Maori honour called for blood vengeance. The chiefs of Rua-tahuna met to consider ways and means ; and the council of Te Puhi-o-Mata-atua said, "Let us raise the war-axe, for this is an evil thing, a jeering at the dead ; we will give them live men to jeer at." Then the army of Tuhoe land went forth from the vale of Rua-tahuna, under the chiefs Tui-ringa, Taitua, Koroki, Tipihau, Te Hiko-o-te-rangi, Moko-nui-a-rangi, Te Whare-kotua, Poho-korua, Te Purewa, Te Umu-ariki III., Tangata-iti, Moko-haere-wa, and Taua. About this time trouble commenced for Ngati-Ruapani, of Waikare-moana. Tuhoe assaulted and took the walled pa of Whakaari, situated on a little headland near Matuahu. The garrison fled in their canoes across the lake to Puke-huia Pa, and Tuhoe at once set about hewing out canoes in which to follow their retreating enemies ; for to be canoeless at Waikare is about equal to being at sea without a vessel, owing to the many cliffs on the lake coast, and the generally precipitous nature of the surrounding country. They made two canoes, which were named respectively Roimata-nui (the abundant tears) and Ruha-nui (the great weariness). The Tuhoe chiefs said, "Let us make a night attack, that none may escape," and this was agreed to. In the darkness of night the force crossed the black waters of Whanganui to the attack on Puke-huia. Te Hiko said, "*Ko au, manu oho atu tenei,*" (I am the early bird of the morning). Te Rangi-pumamao uttered this saying, "*Ko au, ko te tangata i aitia mo te ata hapara,*" (I am the man created for the dawn).

Again the Ruapani fell, and Tuhoe took Puke-huia as they had taken Whakaari. The lake chiefs killed at these two fights were Rangaranga, Tauihu-kahoroa, Moko-ha, and Tu-taua, together with many men of lesser rank. Peace was once more established between these tribes by the raising of the *tatau pounamu,* the "jade door" which closes on war and strife.

Some time after this the Ngati-Kahu-ngunu Tribe, of the Wairoa and adjacent districts, elected to march on Waikare for the purpose of assisting the Ruapani to attack Tuhoe, when an event occurred which gave them sufficient employment nearer home ; for the tribes of Tuhoe land had arisen, and, dividing into two parties, prepared to square accounts, or to open new ones, with the "Children of the Rising Sun." The first detachment went as far as Te Tutira, where they attacked the Kahu-ngunu. This left the Ure-wera and Ngati-Hine-kura† Hapus of Tuhoe, who held land on the western

---

* The *tatau pounamu* is an expression used by the Tuhoe people to denote a formal and enduring peace ; it is peculiar to their dialect.

† Not to be confounded with Ngati-Hine-kura, a hapu of Ngati-Ruapani, a much more ancient hapu.

44

shore of Waikare-moana, comparatively defenceless, as most of the fighting men were away with the war-party of Tuhoe in the Mohaka country. Such a delightful opportunity was by no means to be discarded, therefore Te Horehore and Te Ariki, of Ruapani, took full advantage of it by falling upon the defenceless women, children, and non-combatants of Tuhoe. Many were killed at small isolated *kaingas*—at Te Maire, Tapuae-nui, and elsewhere—in twos and threes, but the bulk of the people were living at Tikitiki, opposite the Mokau Inlet. Here they had cultivations, and a small pa defended with palisades, but no earthworks. Many lived in Te Ana-o-tikitiki, a cave or rock shelter on the western side of the promontory. The Ruapani surprised these people and slew a great number of them, and it is said that the cave was full of dead people. They also threw many bodies into the water, from which act the place and slaughter takes the name of Wai-kotero. Some of the survivors at once started for the south, in order to overtake and bring back the band of Tuhoe who were having, doubtless, an interesting time with the Kahu-ngunu of Mohaka and Te Tutira. On learning of the tribal disaster at Tikitiki, the *ope*, or company, at once renounced the joys of invasion, and, marching by inland tracks to Wai-o-paoa, they skirted the eastern shore of Wairau-moana, arriving opposite Nga-whakarara Isle about half an hour after the fall of that historic pa, but in good time to join in the pursuit, which they did with the fine zest of the Maori of old.

The majority of the Tikitiki refugees, however, fled to Ruatahuna, there to relate their woes to sympathizing friends. In the meantime the second detachment of the great Tuhoe East Coast expedition had left Rua-takuna under the chiefs Tipihau, Koroki, Te Rangi-pu-mamao, Te Ika-poto, Te Hokotahi, Te Pou-whenua, Hautu, Waiari, Piki, and Waikato, together with a company of Ngai-Te-Rangi-ao-rere, a hapu of Te Arawa Tribe, under Te Awe-kotuku, Te Ika-tarewa, and Mataka, numbering, all told, nearly eight hundred men. As this formidable army was ascending the Huia-rau Range, the *mata-taua* (scouts) met two men of Ngati-Ruapani at Poututu, who were going to Manawa-ru to fetch away six of their tribesmen, who were living at that place. As the Tuhoe scouts met them one remarked, " *Kua mate a Waikare.*"* Te Ika-poto asked, " What is the sign?" The old scout replied, "*Inahoki te hahana o te kanohi o te tangata nei*" (Behold the glow in the face of the man)! However, the two men were allowed to proceed, and when the Tuhoe reached Te Pakura they there met the survivors of Tikitiki, who said, " Waikare has fallen ; nothing remains but the drifting waters."

Then the army of Tuhoe rose in wrath and grief, travelling quickly to avenge their slain tribesmen of Tikitiki. They found some of their enemies living at Te Maire, whom they attacked at Whaka-

* " Waikare has fallen." " Our people at Waikare have been killed," or disaster has overtaken them.

komuka with commendable alacrity, killing Toko and his wife. When they reached Whanganui, they could find no canoes in which to cross the lake, for the Ruapani had taken or concealed them all. After a long search they came upon the famous and historic canoe "Hine-waho," which had been dismantled and much damaged by her owners, so as to be unserviceable to the men of Tuhoe. But a great and wonderful treasure had just been acquired by the warriors of Tuhoe land in the form of a few European axes, though the gun was as yet unknown to them. So these new tools were placed in the hands of the Arawa contingent, who were probably the more skilful in the various arts pertaining to the Maori canoe. Under the able direction of Te Awe-kotuku, the big canoe was soon in good order, and the *taua* was quickly in camp on the eastern shore of the lake and preparing to attack Ngati-Ruapani, who had retired to the island forts of Nga-whakarara and Motu-ngarara and to Nga Whatu-a-Tama. The last two of these pas afforded but scant opportunity for the display of Tuhoe skill or courage in war, but the battle at Nga-whakarara was fierce and prolonged. The hapus of Tuhoe who engaged in the storming of Nga-whakarara were Hine-kura, Ngai-Te-Riu, Ngai-Tumatawhero, Ngai-Tawaki, Tama-kai-moana, and Te Ure-wera, the latter hapu being the descendants of Mura-kareke. Having seized canoes belonging to their enemies, the war-party of Tuhoe, with the little band of Arawa allies, crossed the stretch of the lake separating the island pa from the mainland and made a simultaneous attack on both sides of the pa. The *taua* (war-party) were not inclined to linger by the wayside, and made so fierce an attack that Ngati-Ruapani elected to leave for pastures new. It seemed to them that Nga-whakarara was an excellent place to migrate from. So they took to their canoes and indulged in some record paddling for the mainland—that is, with the exception of a goodly number who stayed behind to furnish a fair repast for their cannibal enemies. Some of the refugees made by canoe for the Straits of Manaia, and struck lustily out for One-poto, on the eastern shore. These were pursued by Tuhoe in their canoes, and, as they overhauled several of the enemy's vessels, a series of small naval engagements took place, in which the sorrowing "children" of Ruapani would appear to have got decidedly the worst of it; the survivors landing at One-poto, where they abandoned their canoes and fled to Te Wairoa. Horu, the *tohunga* of Ngati-Ruapani, was killed by the pursuers on the little ridge above the beach between One-poto and Te Kowhai Point, the spot where Herrick's Redoubt was built in after years. At this place, also, Tuhoe built a pa, known as Te Pou-o-tu-matawhero, for the purpose of holding the Ruapani in check.

Those of the island garrison who landed on the mainland opposite Nga-whakarara were also pursued by the vengeful Tuhoe, who landed almost at the same time as the defeated islesmen. Just as the two parties were landing, a strong body of men was observed coming rapidly along the lake-side from the south. This was the southern war-party of Tuhoe, who had been fighting the Kahu-ngunu at Te Tutira and elsewhere, and, recalled by messengers from Tikitiki, travelled by forced marches to Wairau-moana, arriving just too late to take part in the attack on the island fort, but at once joined in the pursuit of the flying Ruapani. Tuhoe, their forces now being combined, chased the unhappy enemy around the shore of the lake. Pare-tawai was killed by Tipihau just opposite the island fort.* Tuhoe killed as they went (*patu haere*), and did not halt until reaching Whakamaro, down the Waikare-taheke River, where a force of Ngati-Kahu-ngunu had collected to assist the lake hapus, with whom they are connected.

The Tuhoe *ope* appears to have remained for some time in this neighbourhood, and lost the two chiefs Hape and Te Ohinu, the latter a younger brother to Waikato. They were killed by the Kahu-ngunu at Tauranga-koau on a frosty morning, as they were lying in a sunny spot to warm themselves.

After this, Tuhoe marched to attack the pas of Pohatu-nui and Pa-nui, by which time they had obtained a wondrous ally in the shape of a *kope* or old-fashioned horse-pistol, called Marama-atea. This *ope* attacked Ngati-Kahu-ngunu outside their pa and killed one man, upon which Kahu-ngunu retired into the pa. Tuhoe then made a sham retreat, appearing to fly in confusion, but the warriors fell aside one by one and concealed themselves in the brush. This was to induce their enemies to follow them into the ambush prepared, which they did (*kua hara mai he hurahura-kokoti*). As they followed the retreating Tuhoe, one of their number, who was in advance, was attacked and slain by Ruru, who took the dead man's *huata*. or spear, and personated him for some time, to delude the luckless Kahu-ngunu, who were defeated by the ambush.

The scouts of Tuhoe entered Pohatu-nui Pa under cover of night to reconnoitre the position. Te Aukihi-ngarae, who had the *kope*, fired it off as a signal to the *ope* without; these now rushed the pa, which fell to them. The chiefs of Ngati-Ruapani killed in the above fights were Whatawhata, Rangaranga, and Te Karaka. Te Ariki escaped, but was captured after a long chase and slain; Tirawhi was enslaved. "*Te rahui kawau ki roto o Wairau*"† is an expression applied to the refugees of Nga-whakarara by Tuhoe, on account of the manner in which they flew from place to place.

The Arawa allies now returned home. As they left Rua-tahuna, Te Purewa, of Tuhoe, said, "Return to your homes; but, lest you be assailed by hunger, do you return by way of Whirinaki, and help yourselves to my potato-heap (*pu taewa*) at that place," the *pu taewa* being the people of Ngati-Whare, Ngati-Haka (Patu-heuheu), and Ngati-Manawa, who were living in that valley. This is the origin of the famous *pepeha*, or "saying," for Ngati-Whare, "*Te pu taewa a Te Purewa*," which same it is well to expunge from one's

* The chief Karetao was also killed, fights occurring at Te Upoko-o-te-ao and Tutae-maro.
† "The flock of shags within Wairau."

whakatauki, or proverbial sayings, when dwelling within the classic vale of the Great Cañon of Toi.

Te Arawa were not slow in taking the hint, and attacked these people at Manga-kino, just below Umu-rakau. Te Rua-ngaio, chief of Patu-heuheu, was killed, and many others were led prisoners by the Arawa to the lake district, some of whom were released by their captors after Christianity had gained a hold on their heathen minds. One of these, Whare-kauri, still lives at Whirinaki.

After this crushing defeat Ngati-Ruapani remained peacefully quiet for some time; but Ranga-ika and his brother chiefs were dark in their hearts towards the tribes of Tuhoe land, and cast about for a plan by which they might obtain *utu,* or payment, for their reverses. And Mokoa uttered the ancient proverb, "*Me ai ki te hua o te rengarenga, me whakapakari ki te hua o te kawariki*" [Create them (men) from the fruit of the rengarenga (evening primrose), and mature them by the fruit of the kawariki (a plant)].

So it fell out that certain *kara* were sent to the Kahu-ngunu of the coast lands, which *kara* were tokens sent by one tribe to another by which they ask assistance to attack an enemy. Closely allied to this *kara* is the *tiwha,* which denotes a similar request for assistance, and may be a material token or merely a hint conveyed in a song. Should a party of people go forth to visit a relative dwelling with another tribe, and should that relative send or present to them a basket of cooked *kumara* or *taro,* and should they find a stone among that food, then is it clear to them that the stone is a *tiwha,* and by it they are silently asked to arise and attack their hosts. Such is the material *tiwha.*

When Te Mai-taranui, of Tuhoe, went north to ask the aid of certain tribes in attacking the Wairoa and Mahia people, he conveyed his meaning to them by means of a song, which he sang to the chiefs of the different tribes in succession--to Te Waru, to Tu-te-rangi-anini, and to Pomare. This song was a *tiwha.*

However, the *kara* was accepted by the Wairoa tribes, who raised a band of warriors and marched to Waikare-moana, where they joined forces with Ngati-Ruapani, the combined hapus being led by the chiefs Ranga-ika, Waiho, Puahi, Toki-whati, and Te Rangi-paea. Tuhoe were not slow in taking up the challenge, and a great fight took place between the opposing parties on the rugged, boulder-strewn beach at Te Ana-o-tawa, a cave which is situated at the base of Te Ahi-titi cliff, close to Te Wha-ngaromanga. Then was heard the clash of weapons as men fought with the old-time arms of the Maori, and the death-cry of many a warrior rose high above the roar of Hau-mapuhia. About fifty men of the coast and lake tribes fell here, including the chiefs Waiho, Puahi, and Mahia.

Ranga-ika, as he saw his best fighting men fall around him, and others flying from the enemy, realised that the battle was indeed lost to him, and that the fighting Tuhoe were again victors. Then came upon him the sickening dread which men feel when they stand face to face with a fearful death, and the excitement is off. His throat was dry and hot and the flow of saliva ceased—an evil omen. Stooping down, he lifted in his hollowed hands cool water from the lake-side, and crying in a strange, hard voice, "*Ka maroke te kaki ; kua mate ! kua mate ! kua mate !* (the throat is dry ; it is death ! it is death ! it is death !) he drank the water. For it was a sign from the gods ; it was a *miti aitua* (an evil omen).

Turning to the cliff of Ahi-titi, which rose above him, he clambered up the rocky ledge which slants upward from Te Ana-o-tawa, and so escaped into the forest above, while a fresh band of Tuhoe, who were now approaching the battle-ground in canoes, lay on their paddles off-shore at Nga Hoe-a-Kupe and sang a deafening *puha*, or jeering song.

Tuhoe, now determined to hold their own at Waikare-moana, built the pas Waimori, Te Waiwai, and Pa-pouaru, and proceeded to camp on the lands as well as harry their unhappy neighbours. Many fights occurred at Te Wairoa, Te Putere, Mohaka, Tutira, Maunga-haruru, Wairau, and Heretaunga. The war became a succession of skirmishes and desultory fights of no magnitude, the result being that Waikare-moana was practically abandoned by Ngati-Ruapani, only the *taha-rua* remaining—that is, those who were related to both sides.

The long-suffering tribes of the lakes and coast then organized an expedition to avenge their defeats, and drive the Tuhoe from the eastern slopes of the great Huia-rau Range.

MOHAKA'S RAID ON TUHOE LAND.

Mohaka was a priest or *tohunga* of the Ngati-Kahu-ngunu Tribe, and held strange powers of life and death, for he was the medium (*kauwaka*) of the god Po-tuatini, which *atua* some call Tu-nui-a-te-ika. He was also a seer (*matakite*), by aid of which wondrous power he could foretell events. It is not given to the multitude to possess this strange faculty.

So the army of Kahu-ngunu arose, four hundred strong, and prepared to scale Huia-rau and attack the men of Tuhoe, who ever lived in scattered *kaingas* and small pas among their rugged forest ranges. And the priest Mohaka prepared to enter into the sleep during which the message or decision of an *atua* is given. So the *tohunga* slept, and his god spoke, saying, "There are two *papa* (or signs) for this war-party — the *rakau-tu-tahi* (the solitary tree) and the *urukehu* (light-haired one). When you capture the *urukehu* do not kill him, but simply degrade him. If you do this, and also see the *rakau-papa*, then shall Rua-tahuna be yours, and Tuhoe will fall ; but should you slay the *urukehu*, then the anger of the gods will descend upon you, and you will be seen scrambling away on all-fours (*Ka haere peke wha koutou*)." Then the *atua* uttered these words,—

Ka noho au i to whenua	I will dwell in thy land
Uki, uki, tau-e !	Generation after generation, years untold.

Then Mohaka the priest awoke from his sacred sleep and returned to this world. And he explained to the warriors the message of his oracle, or god : " There are two signs or tokens (*papa*) for our expedition, one is a tree token (*rakau papa*), and the other a human token (*tangata papa*). The human token is an *urukehu*, a fair-haired man, and should we find this man we must not slay him, but only degrade him before men. Then we must seek the tree token or sign— the lone tree (*rakau-tu-tahi*). If this is found, then shall the word of the gods be fulfilled, and Rua-tahuna shall fall. But if the word of the *atua* be trampled upon, and ye slay the *urukehu*, then shall the children of the rising sun crawl away on all-fours like beaten dogs."

The *papa* herein mentioned is an object, person, or bird seen by, or disclosed to, the priest who is the medium of an *atua's* prophecy. If this certain object be seen, or killed, or caught according to the supernatural direction, then shall that war party be successful, and glory in much slaughter. In the above *matakite* (or vision) are two such *papa*, the man *papa* and the tree *papa*. *Urukehu* is the term applied to the singular and ancient type seen among the Maori people, whose peculiarities consist of a very light-coloured complexion, as that of an octoroon, and red or light-coloured hair. They have ever been numerous among the Tuhoe tribes, and would appear to be the lingering but persistent remnant of some remote archaic type. There are many such singular examples of *matakite* on record here, such as the *kawau papa* for the great battle of Puke-kai-kaahu, between Tuhoe and the Arawa of Rotorua, which was fought out on the shores of the Rere-whakaitu Lake. A strange legend this, inasmuch as the kawau (cormorant) transformed itself into a pigeon, in which form it was much easier to approach and kill. As also Te Hiahia, the *waka papa* or canoe *papa* of the *atua* Te Rehu-o-tainui, which decided the day for Tuhoe when they attacked the warriors of Taupo-moana at Orona, in order to avenge the *kanohi kitea* (incursion) of Tai-hakoa at Rua-tahuna ; that is to say, it did so in conjunction with the man *papa*, the red-cloaked Te Kiore of Ngati-Tuwharetoa. For the men of the inland sea went down to Hades amid the thundering chorus of Tuhoe :—

Ko wai te waka-e !	Which is the canoe, eh ?
Ko Te Hiahia te waka-e !	Te Hiahia is the canoe, eh !
Me he peke mai a Te Kiore	If Te Kiore shall spring
Ki runga ki nga taumata o Uru-kapua ra,	Above to the summits of Uru-kapua,
Ki reira tirotiro ai.	Then shall we see.

And, again, there was—— *Kati !* We will now cease, for the trail is a long one, and the four hundred warriors of Mohaka have passed through the sacred *wai-tuua* ceremony, and are eager to break camp and lift the Huia-rau trail for Tuhoe land, though we shall see that they were still more eager to return.

The East Coast raiders marched by the Orangi-tutae-tutu Stream, falling into the Whanganui arm, thence across Huia-rau, and attacked the scattered Tuhoe *kaingas* of Te Kaha (near Te Wai-iti), Tarewa, Te Hinau, and Mauri-awhe, killing people at all these places. They then built a pa at Manawa-ru and prepared to fulfil the *matakite* (or vision) of Mohaka, and conquer Tuhoe land. So eager were they to commence this contract that they did not wait to properly finish their pa, but sallied forth to wipe out the descendants of Toi and Tuhoe-potiki, of Awa and Tawhaki. This time they attacked Rae-whenua, where the Ure-wera were defeated, and where, to the great joy of the Kahu-ngunu Tribe, they found an *urukehu*, one Matangaua of Ngai-Te-Riu, hapu of Tuhoe. Here was the first *papa*. The survivors of this fight, including the fair-haired Matangaua, foredoomed to degradation by the god Po-tuatini, fled in dismay. The *urukehu* and Te Kaho ran together, and the latter escaped, but the hapless Matangaua was overtaken and captured beneath a lone totara-tree, which stood in the centre of a clearing. Here was the *rakau-tu-tahi*, the lone tree *papa* of the *matakite*, and the hearts of Kahu-ngunu were glad within them.

Here was an opportunity for the descendants of that old *tangata kai paawe* (wandering idler), known to fame as Tamatea-kai-haumi, to achieve greatness, and send their names ringing down to future ages as the bold conquerors of Tuhoe land. But the gods who live for ever had ordained otherwise, and, like the children of Houmea of old, who trampled upon the sacred *aho*, the sons of Kahu-ngunu broke the unwritten law, and so went down to the Reinga (to Hades).

They killed him. Matangaua, the *urukehu* of the prophecy, was slain by those who should have saved his life as the most valuable on earth.[*] Cast aside were the teachings of Mohaka, the *tohunga*, and it is said that the man who slew Matangaua was a *tangata kopu rua*—that is, he had a friendly feeling towards Tuhoe, and so killed the *urukehu* to save him from eternal degradation—to himself and his descendants. If the act which had been commanded by the *atua* had been carried out on the body of Matangaua, it would have so weakened (*whakaeo*) the tribal *mana* and prowess, that the banner of Kahu-ngunu (had they possessed such a thing) would ere long have waved over the earthworks of Ruatahuna-paku-kore.

Tuhoe now collected all the available fighting men in the district, and in a few days a formidable force of Te Ure-wera, Ngati-Rongo-karae, Ngai-Tawhaki, Tama-kai-moana, and Warahoe had met at Rua-tahuna. These Warahoe are not a Tuhoe hapu, but a division of Ngati-Awa. The name of the hapu is said to be derived from that of a stream, and the saying is, "*Ko Warahoe te awa, ko Warahoe te tangata*" (Warahoe is the river, Warahoe are the people also). And again, "*Ka urukehu te tangata, kua kiia no Warahoe*" (light-haired people are said to be of Warahoe). These people were attacked by Ngati-Awa and driven to Taupo, from which place they

---

were again expelled, and they were eventually allowed to settle at
Rua-tahuna. After a sojourn of two or three generations at that
place they went to Te Whaiti-nui-a-Toi, where they may now be
found, living among the Ngati-Whare at Te Murumurunga.

This force attacked Ngati-Kahu-ngunu at daybreak, and suc-
ceeded in defeating them, killing the chiefs Momo-kore, Tautaua,
and Pouheni. The next day another engagement took place, and
another leading man of the Kahu-ngunu was slain. The invaders
fled to the shelter of their fort at Manawa-ru, which, three days
after, was assaulted by Ngati-Tawhaki and Tama-kai-moana, who
killed the chief Poututu and two others of rank. The invading
force now fled under cover of night, carrying the body of Poututu
with them on a litter, up the Rua-tahuna Creek towards Huia-rau.
The pursuing Tuhoe came up to the flying *ope* at the junction of the
Moetere and Ngutuwera Streams, and a merry picnic ensued.
And this is how that placed obtained the name of Poututu. As
for Kahu-ngunu, *kua haere peke wha ratou* (they had gone off on all-
fours). And thus ended the great invasion of Tuhoe land.

The next and final link in the long chain of battles, murders,
ambuscades, surprises, and repasts, which comprised a kind of
profit and loss account between these tribes of Tuhoe land and those
of Waikare-moana, consisted of the expedition of Ngai-Tawhaki and
Tama-kai-moana (formerly known as Ngati-huri) to Kuha-tarewa,
where they fought the enemy, and, when they reached Whakaari
on their return, saw smoke over at Pane-kiri, which induced them to
go over and extinguish that fire, together with the kindlers thereof.

The gleaming camp-fire has burned low down, a chill breeze
comes in from the silent waters, as the Kaumatua ends his long speech
anent the days of old. It remains but to pile on more logs to keep
the fire in, put carefully away the valued note-book containing so
much of the ancient lore of Waikare, conserved in the mysterious
phonographs so puzzling to the "children," and roll ourselves in the
blankets within our sheltering tent. As the Pakeha drifts out upon
the silent waters of Lethe, the murmuring sound of the Kaumatua's
voice comes to him, crooning the old-time ballad of Haere and
Houmea-taumata :—

>Ko te mate o Tautu-porangi
>I haere ra te whanau ki te ngaki i te mate
>Ka tu i te reti, ka ngawha te upoko
>Hoki ana ki te kainga-e-i.

But all *kaingas* are alike now, for the Lethean shore is reached.

Daylight struggles down through the fleecy mantle drawn across
the face of Wairau-moana by Hine-pukohu-rangi, even as a mother
of the Ao-marama (world of light—*i.e.*, every-day world) covers her
sleeping child. Yet a little while and the "White Maid" lifts a
corner of her mantle, and behold! the gallant sun flashes down
upon the forest ranges across the lake. Still lower down the wooded

steeps and crags creeps the sign of the sun-god until it glistens on the placid waters of the sea of Maahu. And tree and rock, and leaflet, each tiny grass-blade by the silent shore, catches the gleaming rays white with frost. Through the breaking mist a lone tree stands clearly out against the white background, and though we know right well that the rocky mass of the Whata-kai-o-Maahu lays unseen below that lone tawai, yet is the effect most strange. Anon the snowy mist rises and drifts across the calm waters, the cheerful notes of the koko are heard trilling forth among the silent "Children of Tane," the bronze-breasted kereru is seen among the rock-nourished kowhai. A pair of swans drift into view from the sheltering mist, looming strangely large as they glide towards the sunny inlet, their family of plump little ones following in their wake. Then, the far off snowy mountains seem to come to us through the vanishing whiteness as if eager to exchange greetings with Ra, the sun-god.

Then lift, O fleecy fog, and raise
The glory of her coming days.

For it is dawn on Wairau-moana, and it is a goodly sight.

Then we descend to the prosaic, for breakfast is ready. So the Kaumatua and the Pakeha seat themselves before the cheery fire and partake of the bounty of the gods. Let no word be said against such a meal, at such a time, in such a camp. For the tea, albeit guiltless of milk, is a beverage for kings, the ship-bread, however hard and unpalatable to those who dwell by city streets, is equal to the oleaginous bacon which hisseth in anger before the fire. Let us draw a veil over this painful scene.

The " children " are here with " Mata-atua,"and the prow of that gallant craft is decked with plumes of the neinei. These are the *puhi* of Mata-atua. So we again embark and go forth upon the waters, coasting along the western shore of Wairau-moana. Very fair and good to look upon is the sea of Wairau, for the clear waters are shimmering brightly in the sunlight, the cliffs and trees and islets are reflected plainly in its calm waters, while above us the blue sky holds but a few snowy fragments of the mantle of Hine. Looking over towards the eastern shore, the scene is a lovely one, so numerous are the inlets, isles, knolls, points, and sandy beaches, with scarce a bare spot, but forest, and forest, and forest. The hills also on that side are small, which permits a fine view of the bush sweeping back to great Pane-kiri. The view from this part is about the finest to be obtained on the lake, which same, as the *Greenville Bulletin* said, "is a big word."

We are now abreast the Korokoro-o-Whaitiri, a delightful little baylet, which would gladden the heart of the genus camper ; and up yon creek is a waterfall most fine to look upon. Of a verity is this a lovely spot, and even the Kaumatua of ours, grim old warrior that he is—even he feels the effect of the scene, and the word comes forth, terse and expressive, " *Me te aroaro tamahine* " ('Tis like a maiden's presence). Then Te Kopuru — another charming little cove. We note that many of these wooded knolls on the points become islands at high water, but the lake is now " low stage "; also, the water is of great clearness, most noticeable where the rocky cliffs slope steeply but evenly down to the lake. The " children " here suggest that this spot be dubbed Te Wai-whakaata-o-Pehi, but the Pakeha objects to the name as savouring of sacrilege. Let us rather leave this realm to the men of old. The Kaumatua remarks that this is a *moana ware* (mean lake), as nothing but the tawai (*Fagus*) is to be seen, there are no *rakau rangatira* (valuable timbers, literally chief-like trees).

Marau and Marau-iti, an inlet dividing into two branches, is now before us—a most beautiful and picturesque spot, with great crags worn into singular forms upon our left, and many signs of ancient occupation on the hillsides, for here the Ngati-pehi dwelt in former days, and here also came that stout old warrior Ropata Wahawaha, of the fighting Ngati-Porou, to seek and slay the Hauhaus of Ngati-pehi and Ngati-Matewai ; but this was in times modern.

From the head of Marau Inlet is an ancient trail to Waiau and Parahaki, and when the lake is " up " a boat can go up the little creek here for nearly a mile. But the " *moana ware* " expression is not good, for the brilliant rimu lights up the sombre tawai forest, and the koromiko and toatoa, neinei and tupakihi trees all tend to relieve the eye. Then on past Te Kopua, named from a pool where the wily duck is taken by the wilier fowler, to Nga Hina-o-Te-Purewa (the grey hairs of Te Purewa), which same is a tawai-tree overhanging the lake, and which is covered with long grey moss, giving it a most venerable appearance. And no wonder his hair turned grey, for a fiercer old paynim never lived, nor a more pronounced cannibal.

Just beyond Te Rata, the stern frontlet of Pane-kiri looms up again across Whare-ama, and then we glide over the placid waters of Te Totara, a lovely little bay, with a grassy slope running round to an inner bay, and before us is the picturesque and sacred isle of Pa-te-kaha. An ancient fort this isle, one of the oldest pas on the lake, but now covered with forest growth, for it has served as a burial-ground this many a year for the sons of Ruapani of old. It also has the distinction of being the largest island on the lake, which is not a " big word."

Now we pass through the little passage between Pa-te-kaha and the mainland, and enter the beautiful inlet of Te Puna with its green slopes reaching back to Puke-hou. Past Wai-haruru, whence ran an old track to Huia-rau and Rua-tahuna, and Te Upoko-o-Hiwera (named from an ancestor of Ruapani) to the sloping beach under Pukehou, where we again go into camp and pitch our tent, though the Kaumatua stoutly maintains that to pitch a tent in fine weather is to ask for rain.

As the white tents arise in this lone spot, and the sun sinks down behind the western ranges, fain would we speak of that scene at Te Puna, looking across the little bay backed by wooded hills ;

but human endurance has its limits, as also human patience. But the Pakeha, who sits by his tent-door on that golden evening, sees not only the scene before him, but those which have passed by long years ago. He sees the ancient land of Maahu, as Hau and Tama of old saw it—sees the lone lands, unknown of man and innocent of human blood. He sees the coming of the ancient people from the shadow-laden fatherland, and knows full well their deeds and strange customs. He sees them multiply in the land, and the coming of war and strife—the smiting of the old-time people by the migrants of " Horouta."* He sees plainly the ancient kaingas by the lakeside, and recognises the men of old as they follow each his strange art. The trees have faded from the ancient forts across the shining waters, the palisades and great carved himu (posts) are again in place as of yore, the warriors are lashing the huahua (rails) and forming in line for the tutu ngarahu (war-dance). The tohunga, clad in sacred maro (girdle), approaches the tuahu (altar) to perform the holy rite of tira ora. The naked mass of bronze-hued warriors leap into life. Hark! It is the hoarse roar of the war-song which booms across the placid waters and echoes among the world-old hills above.

The scene fades away, and then across the waters come the canoes of the men of old, carrying some chief to his last home on the " Sacred Isle." And as they paddle onward they chant an ancient lament for the dead, old as the days of Maui and of Taranga. They come to land as the first stars gleam in the calm waters beside them, the bearers take up the sacred burden, the priest wails forth a weird karakia (prayer), then the procession winds up the hillside and is lost to view in the glooming forest.

Long tails of fog were streaming up the gullies as we boarded " Mata-atua " for her third day's cruise, and the " Sacred Isle " stood out lone and distinct against a sea of mist.

Te Parua-o-Rora (the bowl of Rora) : This point takes its name from a curiously-shaped stone at the base of the cliff. Paraharaha derives its name from a pool of black mud in which the flax-fibre was dyed in former times. As we pass out of Te Puna Bay, a singular effect of sun and fog is noted. The fog lies in a mass about 300ft. above the surface of the lake, and the sun shining through it imparts a beautiful golden hue to the mist beneath. We glide on to Pakinga-hau, a most suitable name for this place, and enter the Straits of Manaia. The mist breaks open and the sun flashes down on the face of the waters, following us quickly along the rocky coast and lighting up the forest above with a cheery gleam—to Te Upoko-o-Hinewai, named from an ancestress of these parts who flourished some twelve generations ago. So we turn our backs on the " Sacred Isle," and go forward over the shining waters of Te Kauanga-o-Manaia to Weka-ku, so called from an ancient member of Ngati-

---

* One of the early canoes, before the time of the fleet.

Waikaremoana – A Peep from a Cave.

Rakaipāka, and once more pass over the broad surface of Waikare-moana. The waters are sparkling in the sunlight, and long streamers of silver mist lie against the wooded ranges under Ngamoko. In a little baylet over at Mokau, a wisp of blue smoke rises slowly from a camp of Natives, who have met here to perform some heathen cere-mony in connection with taking the *tapu* off certain lands. A hail from the waters astern, and the next moment, shorn of her former glory and ancient beyond compare, "Hine-waho" swings past us on her way to the Hauhau camp.

Then the famous Ana-o-Tikitiki, named from a descendant of Kahu-ngunu. The "children" lay us alongside of the historic cave, and we look into the rocky chamber where so many of the women and children of Tuhoe went down to death. Then the equally famous Puke-huia Pa, now covered with forest growth, and no longer containing the fierce warriors of old—to Hau-taruke, a sacred spot in former times, for it was a *toronga atua*, the sacred altar of the gods Haere, Maru, Kahukura, and Rongomai. Here came the priests of old bearing the sacred symbol of the god, a carved stick, which was stuck in the ground, and upon which the *tohunga* kept his hand as he uttered his prayers, and the *atua* would manifest itself by shaking the stick, and so give its decision. Past many ancient settlements, we pull in to the inlet of Whanganui and explore its many bays, so rich in old-time legend. At Tawhiti-nui we listen to the story of that ancestor who, after death, became a *taniwha* in the lake at this spot, though he does not appear to be of the man-eating variety. He simply appears to men, probably for the fun of seeing them run. Thus Te Waiwai, of Ruapani: "I was in my canoe, fishing for maehe at Tawhiti-nui, when I heard a strange sound, and two great waves came rolling in from the lake. Then resounded two loud reports like unto the cannon of the white men. Then I knew that the *taniwha* was angry. Friend! I quickly plucked a hair from my head and cast it into the water, at the same time uttering a *karakia* to render the demon harmless *(hei whakaeo i te taniwha)* and to calm the waters."

Here is Wai-mori Pa, at the head of the little bay where the Opu-ruahine Stream enters the lake, a picturesque spot and the scene of many an old-time fight in the days gone by; and the old battle-ground of Te Ana-putaputa, where the descendants of Ruapani went down before the "Children of the Mist." Here, also, a hun-dred feet from the shore, are strong springs of water, ice-cold, gush-ing rapidly up from the lake-bottom. As we look over the side of the boat we can see the rush of the spring water from the lake-bed many fathoms beneath the keel of "Mata-atua." Here we appear to be in a small land-locked lake surrounded by high ranges, as the entrance is concealed by a projecting point; but on rounding this point we see before us the broad stretch of glassy waters reaching to far-away One-poto and Nga Hoe-o-kupe. On either side, the over-hanging trees are clearly reflected in the calm waters, presenting a singular and lovely sight. As we pull on down the rugged coast-line towards Mokau, it is most interesting to note the strange irregularity in the strata of the rocky cliffs, for here they are horizontal, and a hundred yards further are vertical, a little further and they again have a heavy list to port, if this scientific term be allowable. On past the houses of the old-time people, and the bush-covered and silent forts of Pa Pouaru and Te Waiwai, where the Kaumatua breaks forth into a *tangi* for the ancient homes of his tribe and those who held them. Then he descends to the practical, nineteenth-century view of matters, as he says, " Should it happen that the forts of Whakaari and Puke-huia were to be at war with each other now, I think that the men thereof would be able to fight without leaving the pas, for a bullet will travel a hundred miles—or is it a hundred yards?"

The beautiful Bay of Mokau is now before us, and we glide over the sunny waters towards the entrance of the stream of the same name. Here the prow of " Mata-atua," the much-travelled, is brought to land, and, while certain of the " children" remain as boat-guard, the rest of our party wend their way up the stream to obtain a view of the falls. A fine sight are these same falls, for they are situated in a most rugged and picturesque gulch. From a ravine about 25ft. in width, the mass of waters fall over a cliff about a hundred feet high, not falling directly into the great pool below, but on to a huge projection of the cliff, a semi-circular abutment—which has the effect of spreading the falling waters out into a great white expanse of foam. The steep forest-clad ranges, rising abruptly from the water's edge, the bush-clothed cliffs, and singular strata all combine to present a most striking effect. To the right is a cave, by which a person may pass behind the great mass of falling waters ; and on the left is a smaller fall, almost concealed by the dense timber-growth.

Then, after duly admiring this fine scene, we wend our way back to our gallant craft, but decide to take the creek-bed instead of following the trail, which runs along the sideling above. So we pull off our shoes and start gaily down the shingle bed ; but it is sad to relate that our aboriginal guide was left far behind by the Pakeha in tramping over the stony channel. Verily an unworthy descendant of Rakaipaka this same guide—a fine fellow to join a war-party bent on scaling rocky Huia-rau !

So we drift out again into the gleam of sunny Waikare, and down along the abrupt coast-line to Whakaari the renowned—Whakaari of Mokoa and many another bold warrior of the long ago. This historic pa is situated on a little point in a small semi-circular baylet, with bush hills rising behind—a truly beautiful spot in summer days. Near by is the promontory of Matuahu, a striking land-mark, and where the sons of the soil closed in battle with the invading Pakeha in the troublous days of the sixties.

But the commissariat of " Mata-atua " has now grown somewhat slim, and we therefore decide to pull across to One-poto, and there camp. On landing we find the historic canoe " Hine-waho " drawn

Fall on the Mokau River, Waikaremoana

up on the beach, the lone survivor of the fleet of former days which floated upon the waters of Waikare. She is about 60ft. in length, and presents a poor appearance, for the *rauawa* (top-sides) have long gone, and her wounds are many. So we camp down by the old pa Te Pou-o-tu-mata-whero, which is near unto the walls of Herrick's Redoubt, on the spur above Te Kowhai.

The next day we elect to remain on shore, for a strong wind is blowing, and the Kaumatua as usual brings forth many proverbs and wise old saws to prove that it is not well to provoke the god, Tawhirimatea. So we stay by the land, and go out to look upon the homes of the old-time people and view the battle-grounds of old. " Friend," says the Kaumatua, " If we had a Pakeha canoe, what you call a buggy, we would go to Te Wairoa and look upon the lands of Te Tauira, but we shall view the Cave of Tawa, and look at that battle-ground where the Ruapani sank in death." So the Kaumatua and the Pakeha clamber round the rugged shore past Te Whangaromanga, and look upon the waters rushing down into unknown caverns below, to the cliffs of Ahititi, with their singular strata, symmetrically fissured, as if some Titan of old had amused himself by arranging here Cyclopean walls and buttresses, and strange overhanging table-rocks, the softer strata being worn out by the winds of many centuries; with caves and holes and strange chasms of uncanny aspect, an ideal spot for the cragsman. And we look upon Nga Hoe-o-kupe (the paddles of Kupe), which consist of a rock standing out in the water, a rock with singular vertical fissures dividing it. And here the Kaumatua points out Te Waka-o-Kupe (the canoe of Kupe), a sunken rock, which he declares is the canoe of that old sea-rover ; and on a calm day you look down through the clear water and see the men of old seated therein, with paddles in their hands, as if waiting for their old commander to return from his wondrous voyages to far-away lands, and then they will once more go forth upon the dark ocean as of old, and follow the setting sun to his mysterious cave, and conquer the dread demons of the sea, by potent spells of fearful import, and sail down to unknown lands which lie beyond the sky, and see the strange men and strange products thereof, and camp again with Turi at Rangi-tahua, and meet the rising sun on the edge of the world, and lift again the old landmarks at Karotonga and Tawhiti ; and the golden days of the brave old world-finders shall return at last.

Then the old warrior goes on to tell of wars of old and many strange things which happened in former times, and also explained that Nga Hoe-o-Kupe is a rock possessed of great *mana*, for should any one strike it, the wind will at once change. Then we enter Te Ana-o-Tawa at the base of the great cliff, and in that spot, where the men of Ranga-ika strove against the warriors of Tuhoe land, the Kaumatua once more opens forth, and describes that Homeric struggle in vivid language and with appropriate gesture " And that was how we slew the Ruapani in the old days. Then we and the *taharua* (people related to both sides) held these lands, which are now lone and deserted of man. In my young days, when I lived on the further shore, I could see that the hill above One-poto was covered with large *whares* (houses), and the great *himu* (posts) were standing. And in those days it was that I heard Tutaua, the log-demon, singing in the darkness of night. This Tutaua was a *tipua* (a spirit, a demon) in the form of a totara tree or log, which was placed in the lake by Hau-mapuhia, son of Maahu. This demon log was ever floating on the surface of the lake, ever drifting across the waters from place to place; and it sang strange songs as it floated upon the dark waters of night-ridden Waikare—songs of strange import were they. The people living upon the lake-shores would often hear these plaintive songs afar off. At such a time the old people would say, ' *Ko Tutaua c waiata haere ana* ' (It is Tutaua, singing as it goes). I myself heard it at Rerewha, in my young days, singing in a strange voice, like the wind whistling. If the log drifted ashore, and should any person break or cut a piece of wood off it, Behold! the next morn that log demon had disappeared. Tutaua drifted away out of the lake through the outlet at Te Wharawhara when I was a lad—drifted away, singing as it went."

There is no holding him now ; for the old fighter is once again started on the beloved subject of the men of yore—their deeds, evil and otherwise, in the world of light—and tale after tale comes of wars and sieges and priestly craft, as the Kaumatua drifts back over the stormy sea of his adventurous life, and greets again his old comrades of the war-path, and again takes his place at the camp-fires whose ashes have been cold tor half a century.

Then we drift away from the historic Cave of Tawa and go back through the flying spray, with the roar of Hau-mapuhia in our ears, to the hill Raekahu, which stands above One-poto. And we ascend that hill to look upon the lands below, and the little lakelets of Nga-whakatutu, Wherowhero, and Te Kiri-o-pupai. And here we stay awhile and observe that fine scene—

A ROUGH DAY ON WAIKARE-MOANA.

For a strong wind swoops down through the mountain-passes and inlets of Waikare, lashing the waters into foam. The white-crested waves are surging across the troubled lake, and break in wrath on the rock-bound shore at Te Wha-ngaromanga, on the great buttress of Ahititi. The spray is flying over Te Taunga-a-tara and Nga Hoe-o-Kupe, and drifts across the divide by the narrow outlet, for the "Sea of the Rippling Waters" is awrath and pounds heavily on the imprisoning cliffs as if eager to be free.

To the west, the dark clouds are glooming over distant Huia-rau, and driving down the rugged defiles which open out on Wha-nganui and Wairau. Then a struggle ensues between the storm and the westering sun for the mastery of Waikare, but the gallant sun-god triumphs, the rain-laden mists sag downwards and possess the inner arms of the "Star Lake," while the vapours above are white and fleecy beneath the conquering rays, and far away across the

Near the Outlet, Waikaremoana.

tossing waters a broad stream of gleaming silver stretches even unto Nga Whatu-a-tama and Te Upoko-o-Kahu-ngunu.

As we view the expanse of angry, surging waters, and listen to the hoarse roar of the white surf, it is hard to believe that this is but an inland lake, and not an inlet of the great ocean, with the swell setting in from the far Pacific.

Anon the sunlit mist settles down over the ranges of the west, obscuring the frontier of Tuhoe land — the giant Huia-rau. The white scud flies athwart the darkening ranges above Whanganui, the wind moans through the sturdy rock-nursed beeches, and among the weird cliffs of Ngamoko and Pane-kiri, but the bright sun flashes upon the heaving waters of Waikare, and the heart of man is glad.

Such is the view from Raekahu on a stormy day. But the following day was fine and calm, as we explored the wonders of the rugged cliffs around Pane-kiri and Awaawa-roa. The road now in process of formation from One-poto around the lake to Aniwaniwa is a delightful walk, and from it is obtained a fine view of the lake. And here on either side are strange caves and holes, yawning chasms of unknown depth, huge galleries running far into the range, and overhanging masses of rock. Yonder stands a huge splinter of a hundred tons or so, on the summit of which a great tree has perched ; here is a rock, 20ft. in length, under which a rata has grown and lifted the great mass up bodily. Below us lies the now placid lake, and far away the snowy mountains rise sharply against the sky-line. The sun sinks down on distant Huia-rau, and lights up the great Pane-kiri Bluff, as we wend our way homewards. A long streak of golden light glitters across the calm waters and follows us as we go forward. The shadows on either side deepen into purple, and from far away across the gleaming lake comes the sound of a heathen song, as of Hau-ma-puhia or Tutaua, "singing as it goes." The canoe of the singers glides across the golden stream and is lost in the gloom of Pane-kiri, and then, with that glorious light gleaming on golden mountains and glittering waters, and the voices of the night around us, we go down into the darkening valley below.

WAIKARE-ITI.

Our next trip is to Waikare-iti Lake, which lies east of the Aniwaniwa Stream, and is about 500ft. higher than Waikare-moana. So one fine morning found us again setting forth and pulling down the coast-line to the Whanganui-o-parua Inlet. Past the lone Wha-kangaere Rock, another famous *ahi-titi* of former days and Kakata, so named from a sister of the famous chief Te Purewa, though it was no laughing matter for poor Kakata (laughter), for she was drowned here, together with six others, by the upsetting of a canoe. Then to the Hinaki-o-Tutaua, which exists in the form of a rock, but as to what use that cheerful *tipua* could make of a *hinaki* (eel-basket) is un-known ; still it serves a useful purpose, for if a north-west wind is blowing, and one does but stroke the rock with the hand, the wind will at once change to the south, which same is useful information, as during a south wind this side of the lake is sheltered. And Te Heru-o-Hine-pehinga, where doubtless that maiden of yore was wont to prepare her simple toilet, inasmuch as this was a famous place for the heruheru fern, of which combs were made. Along the shore are many signs of ancient occupation, but now *ko te moana anake e tere ana* (there remains nothing but the drifting waters). We now pass cliffs of blue papa, and the effect of the green shrubs and blue cliffs is quite striking ; and the two rocks known as Tuara and Ruatapunui, which stand out in the lake, and many delightful coves and little beaches which make one yearn to camp down for a while. At Kirikiri was a famous *moari*, or swing, in former times, where the young people amused themselves by swinging out and dropping into the deep waters.

We camp at Te Papaki for the night, this place being at the head of the inlet, and, besides being a good camping-ground, is well situated for the advance on Waikare-iti. A boat can pass up the Aniwaniwa Creek here to the first fall, but the big falls of Papakorito are some distance further up.

The next morning sees us ferried across the head of the inlet and landed on the right bank of the creek, from which spot a two-hours' walk up the range brings us to Waikare-iti. On reaching the top of the hill, we descend a small spur for a short distance, and see through the trees before us the calm, silent waters of Waikare-iti. This beautiful lake is surrounded by low hills covered with dense forest, which extends to the water's edge, the branches trailing in the water in many places. There are none of the great cliffs and ranges of the larger lake here ; the scenery is not grand, as is that of Waikare-moana, but it is nevertheless very beautiful, there being many little islands in the lake, all densely clothed with bush. One longs for a canoe at canoeless Waikare-iti to go out and explore those lonely islets, and paddle across the shining waters. We are fortunate in happening upon the one spot on this side, apparently, from which a good view of the lake can be obtained. A great rock juts out some distance into the lake, and on this rock we seat ourselves, disturbing thereby two whio (mountain - duck) which were taking a siesta below. The lake is probably a mile and a half across, but the view of the further shore is almost concealed by the islands, of which there are six—Motu-torotoro, Motu-ngarara, Te Kaha-a-tuwai, Te One-o-tahu, Te Rahui, and another, of which our guide did not know the name. Truly a lovely scene this on such a day, the calm, clear waters glittering in the rays of the sun, the lone, silent waters, surrounded by dense forest, and, in the far distance, the snow-capped peak of Manuaha. There are no signs of ancient cultivation here, as on the shores of the "Star Lake," but this place was occupied by the Ngati-Ruapani Tribe as a place of refuge. When harassed by enemies in their *kaingas* at Waikare-moana they would retreat here and occupy the numerous islands in the lake, drawing their supplies probably from

the surrounding forest, for the diminutive mache is the only fish found in these waters, though wild-fowl were formerly numerous, including the whio, maka, weweia, and kaha, the latter a large bird which nested in the branches trailing into the water on the shore-line. The timbers seen here include the tawari, toatoa, tawai, horoeka, tawhero, parapara or houhou, neinei, miro, papauma, horopito kaponga, and the punui fern, with many others of that beautiful tribe. The outlet from Waikare-iti is by a swift stream, which flows with a heavy fall towards Aniwaniwa Creek.

A weird and silent place is Waikare-iti, with its unexplored isles and great forest ; a most beautiful and unknown spot, but bearing no sign of the presence of man. Verily the Bohemian spirit longs to go and explore those silent islands and search for traces of ancient occupation thereon, from the days when the "children" of Ruapani and the ancient Tauira held these lone lands. But we lack the time to go a canoe-building, so we turn and retrace our way to the camp at Aniwaniwa—at least some of us do ; but our worthy guide stoutly maintains that we are on the wrong trail, and, as we refuse to believe him, he set off on his own sweet way, with the result that he got lost and wandered around the spurs of the range while we are in camp discussing sundry viands furnished by the great harbour of Parua—to wit, *puwha* (edible thistles) and *kakahi* (shell-fish). And in returning we get a fine view of the lower falls of Aniwaniwa through the overhanging forest trees, which same is a truely fine sight, for the mass of foaming waters falls in two great leaps some 60ft. to the stream below. As we are striking camp, we hear a hail from across the inlet, and there behold our lost guide standing on a long sandspit running out into the lake. And as we pull out into the lake he wades out into the water to be picked up, looking very forlorn and comical. So we lay in and take him on board amidst many jeers and jibes from the "children," which some-what annoy the old fellow, inasmuch as he remarks that he never knew so many fools to be contained in one boat—which same is distressing to a fine mind. Be not cast down, O faithful Waiwai! for truly art thou a goodly comrade and a cheerful, when camped in the lone places of the earth. And thou art the man who kept a given word, and turned to help the strangers from across the snowy moun tains when the whole of Ruapani had said, Waikare-iti should not be trodden by the Pakeha. *Kia ora koe* (May you live) !

Then we proceed along the western side of the inlet, so as to complete our traverse of the shore-line of Waikare-moana. Past Te Ana-o-Tuaraia, so called from an ancestor of Nga-Potiki ; and, lest you be surprised at the number of dead trees in the forest on the range above, it is as well to know that they were destroyed by witchcraft, by the Kahu-ngunu people of Te Wairoa, a tribe ever famous for their powers in *makutu*, hence the expression "*Wairoa tapoko rau.*"* And at Whaitiri yonder is a *tipua*, in the form of a

* Wairoa, where hundreds sink.

log, which lies beneath the clear waters, and should that demon be interfered with, then assuredly the whole lake rises in wrath. At Te Wai-a-te-puranga is a strong spring of intensely cold water gushing up from the lake-bed, such as are seen at Te Ana-putaputa —hence the name of this spot. So on past Taumatua, where our Native allies fought the Hauhaus during the last war, and Te Mara-o-te-atua—where we wonder what the gods could possibly have cultivated at such a rocky spot—and the long headland of Matuahu, where the chain is complete, and we drift back across the shining waters to One-poto.

It is the summit of massive Huia-rau again, with the sun sinking in the golden west and the gloom of night settling on the far ocean, for Waikare-moana is far below us now, as we stand on the snow-wrapped peak of the great *ikawhenua* (mountain backbone). We have toiled up the rugged creek-beds of Te Onepu and Wai-horoi-hika, every rock and stone therein covered thick with slippery ice. Long icicles hang from the cliffs on each side like clear stalactites, the great boulders and smooth bluffs are as glass, even the running water is frozen over in many places. So we go forward, barefooted and be-swagged, through ice-cold waters, and toiling carefully up the ice-covered rocks. So slow, indeed, is our progress that the night is falling when we reach the summit, which means that we have been five hours in ascending 2,000ft. Here, then, we proceed to camp, and thus spend our last night in the wilderness. And, while the " children " go on to pitch the tents at Te Pakura, we tarry awhile on the summit to take our last look at the " Star Lake " lying far below, and watch the wondrous glories of the setting sun across the western ranges. For surely it is a noble sight. Towards the west lies a great far-reaching chaos of rugged ranges, valleys, and peaks. Here are many noted mountains of Tuhoe land ; here is Maro, and Whawharua, and Tara-pounamu, and Te Ranga-a-Ruanuku, and Manawa-ru, and Manu-ruhi, and Nga-heni, and Tawhiu-au, and Te Ihu-o-Awatope, and Panui-o-Rehua, and Tane-atua, and Te Niho-o-Kataka ; and far, far away is the great Pae-roa, and still further the giants which look down on Taupo-nui-a-Tia. A glorious light is on the distant mountains, a golden haze fills the valleys and lingers on the plain-lands, the ranges darken to south and north.

The " children " have broken a trail through the snow, and their camp-fire gleams brightly on the spur below, the sun disappears into that golden fairy-land of the west, as the Kaumatua and the Pakeha take their last look at Waikare-whanaunga-kore, and, turning to the gleaming *kura* (red light), go downward through the snows of Huia-rau, *en route* for the Great Cañon of Toi.

—

APPENDIX I.

—

FIGHTING BETWEEN NATIVE CONTINGENT AND REBEL HAUHAUS.

THE FIGHT AT TE KOPANI, 1865.

THE rugged country around Waikare-moana was a great rendezvous and stronghold of the rebel Natives during the Maori-European war, and troops engaged in the process of clearing the Hauhaus out of this section of country underwent many privations, for the life was a hard one beyond measure.

At Te Kopani, about four miles from One-poto, a severe fight occurred between a force of friendly Natives and rebels. This place is a narrow gully between two hills, and up which the old trail from Te Wairoa to One-poto ran. At the time of this fight, the gully and hills were covered with a dense growth of fern. Some five or six hundred rebels took their stand in this gulch, and proceeded to entrench themselves by sinking rifle-pits, which were skilfully masked by the wily Hauhaus. These were situated on three different slopes, and were not discernible by a person passing up the gully. The friendly chiefs, Ropata Wahawaha, Kopu, and Ihaka Whanga, led a force of five hundred of our Native allies from Te Wairoa against the enemy, and, when marching up the ravine at Te Kopani, they received heavy volleys from both sides, which killed six men and wounded about twenty-five. The friendlies were thus at a great disadvantage, as they were exposed to the fire of an unseen enemy, but the wily aboriginal was not long in seeking cover and replying as well as possible to the enemy's fire. After a period of desultory firing the enemy advanced, and then that gallant old warrior Ihaka Whanga called on his men to charge, and drive back the Hauhaus. But the sons of Kahungunu, never over-distinguished for prowess on the battle-field, declined the sedu - tive offer. Ihaka, however, advanced, and in so doing received a wound in the hip. Having discharged his piece at the enemy, he took another carbine from one of his men, and again fired into the body of Hauhaus, receiving at the same time another wound, which felled the old warrior. His men now rushed forward to recover the body of their chief, which they accomplished, and Ihaka recovered to again meet the enemy on many a future field.

* For information concerning these engagements I am indebted to the kindness of Mr. Tunks, of Windsborough Station.

All this time the enemy had a strong advantage, and many friendlies were killed. Whereupon the European officers (Major Fraser and Captain St. George) conferred with the Native chiefs, and a retreat was decided upon. But Ropata thought that it would be an excellent plan to fire the fern, and so dislodge the enemy. No sooner said than done, and in a few minutes the fire was roaring up the hillsides, creating dense volumes of smoke and driving the enemy from their rifle-pits. The exultant friendlies now took possession of the ridge, and opened a sharp fire on the retreating Hauhaus, who lost heavily in this engagement, and eventually fell back on the wilderness of Waikare-moana, leaving nearly eighty of their dead upon the field. It is certain, however, that this does not represent the enemy's loss, and even now the oncoming Pakeha often finds mouldering skeletons in gully and cave, with probably the remains of a gun by the side thereof.

Colonel Herrick's Expedition against Waikare-moana, 1869.

During the above year it was resolved to despatch a strong force against the rebel strongholds of Waikare-moana. This force consisted of nearly nine hundred men, of whom three hundred and fifty were colonial troops, and the balance made up of friendly Natives of the Ngati-Porou and Ngati-Kahungunu Tribes. The object of this enterprise was to destroy the crops and food-supplies of the hostiles, and to reduce several positions taken up by them at Matuahu and elsewhere on the western shores of the lake. This district had become noted as a refuge and recruiting-ground for rebel leaders, such as Te Kooti, who, raiding down from these secluded ranges on the European settlements, rendered life and property alike insecure on the East Coast.

On arriving at One-poto, a redoubt was erected on a small hill overlooking the lake, the earthworks of which are still standing. Here a long stay was made, and great preparations undertaken for the destruction of Matuahu, the principal stronghold of the hostiles. Two large boats, each 40ft. long, were built, also some metal pontoons, which, with a whale-boat and dingy, comprised a most imposing fleet, by which it was calculated that two hundred men could be landed at one time on the western shore. Matuahu was described as a very strong place, and not to be taken without severe fighting, though it appears that many of the defences were fictitious, and merely intended for show. The intelligent aboriginal also bethought him of discharging at sunset on each day a heavily-loaded gun, the report of which was so magnified by echo that it was thought to proceed from a young cannon.

After this expedition had made preparations for transporting the force across the lake to attack Matuahu, and a start was at last to be made in the great cleaning-out of the lacustrine pas and retreats of the enemy, they rose up one fine morning and retreated on Te Wairoa—whereupon the hostiles crossed the lake to the site of the